ISBN 0-9603120-0-5

Human Body Growth

in the

First Ten Years Of Life

HOWARD V. MEREDITH
University of South Carolina

The State Printing Company
Columbia, South Carolina

Library of Congress Catalog No. 78-63120

ISBN 0-9603120-0-5

Preface

This book provides a series of short chapters on body growth before birth, in infancy, and during childhood. It makes available in concise form, knowledge of interest to (1) the general reader, including parents seeking information on body growth of the human embryo, infant, and child; (2) persons teaching courses on human growth, child development, health education, or related subjects; and (3) individuals preparing for careers in pediatrics, physical education, nursing, physical anthropology, developmental anatomy, orthopedics, child physiology, and other biological specialties.

The emphasis throughout is on factual knowledge—knowledge pertaining to changes with age in external size and shape of the body, effects of maternal cigarette smoking during pregnancy on birth size of offspring, variations in time of emergence and discard of teeth during infancy and childhood, changes during the last century in the body size of children, variations in body size and shape among child populations living in various parts of the world, differences in body growth of children reared in underprivileged and privileged families, and differences in body growth among individuals receiving superior nutritional and health care. The facts presented expand perspective—also point to needs for further research—regarding effects on body growth of variables such as genetic inheritance, altitude and temperature, density of population, food, disease, sanitation, exercise, occupation, and socioeconomic status.

The book draws upon scholarly reports by many investigators. Were it a research monograph, citation of references in the text would be obligatory. Having been written for less specialized use, the book is not encumbered with text citation of references. Instead, the short list of suggested readings at end of each chapter includes one or more publications marked with an asterisk: these publications specify many of the sources used.

Measurement units in the book are principally inches, ounces, and pounds. Millimeters, centimeters, grams, and kilograms are interposed occasionally to help the reader accustomed to thought in terms of inches and pounds make the transition to thought in the metric system.

There are a few technical statements. These are so placed that they can be passed over without losing the gist of the idea under discussion. Anatomical and statistical terms are kept at a minimum. Average and mean are used synonymously.

The book is compact. Readers will want to return to parts of it at later times; to facilitate this, a subject index is supplied.

June 27, 1978

Columbia, S. C.   HVM

CHAPTER I

Growth of the Human Body Before Birth

The gamut of somatic change during prenatal life is vast. It includes replication and obliteration of cells, emanation and resorption of tissues, repositioning and modification of organs, emergence and submergence of external features, and varied trends of alteration in size and form of the head, trunk, and limbs.

Cells elongate, flatten, invaginate, and migrate; they live varying amounts of time—skin cells for a few weeks, blood cells for a few months, nerve cells from when they appear into postnatal life. Organs bulge, loop, bend, convolute, rotate, and tilt. The stomach begins as a swelling of one portion of the digestive tube, passes from this capsular form to a shape resembling a cow's horn, then alters to almost a J configuration. From its early location high in the trunk, it moves to a lower level, rotates clockwise, and tips to an oblique position from left to right.

During the middle third of prenatal life taste receptors become numerous on the tongue, tonsils, palate, and parts of the esophagus; in late prenatal life there is reduction in their number and extent. Many bones develop by increasing in one region and decreasing in another. The mandible (lower jaw) grows in width by bone deposition on its lateral surfaces and bone absorption along its medial surfaces. The shafts of the long bones increase at their peripheries, while at their centers there is bone decrease resulting in formation and gradual enlargement of marrow cavities.

Length of the prenatal period. Human prenatal life begins with the merging of a male sperm and female ovum to form a unicellular zygote; barring abortive termination, it ends with the event of birth. The average duration of this period is 38 weeks. Most liveborn infants are somewhat younger or older than 38 weeks at birth; about 70% vary in age from 36 to 40 weeks, while ages for roughly 98% are spread between 34 and 42 weeks.

Size of the zygote. When ontogeny begins, the human zygote is a spherical cell about 0.005 milligram in weight and 0.14 millimeter in diameter. It would take over 5 million such cells to equal 1 ounce, and about 180 placed in a row to equal 1 inch.

[1]

Single and multiple zygotes and births. Usually a single zygote gives rise to one individual; occasionally it gives rise to two or more individuals designated monozygotic twins, triplets, or quadruplets. Sometimes two individuals (dizygotic twins), or more than two individuals, are produced from almost simultaneous formation of two zygotes.

In the white population of the United States, frequencies of multiple births relative to singleton births are about 1 in 80 for twins, 1 in 9,000 for triplets, and 1 in 570,000 for quadruplets. Multiple births are more frequent in the black population of the United States, occurring about 1 in 70, 1 in 5000, and 1 in 240,000 for twins, triplets, and quadruplets respectively.

About 25% to 35% of liveborn twins are monozygotic, and 65% to 75% dizygotic; monozygotic multiple births are identified postnatally using information from viewing the iris of each eye under two or three magnifications, examining finger and palm prints, and making several reaction tests on blood samples.

The first two weeks of prenatal life. During the first half of the initial week, cell multiplication changes the unicellular zygote into a morula (mulberry-like structure) composed of 16 to 30 cells. There is no increase in overall size: the 20-cell morula is no larger than the 1-cell zygote. In the last half of the first week both cell production and enlargement of the organism occur. At the end of the week, a single flimsy layer of cells surrounds a cavity (trophoblastic layer and vesicle), and in one area of the trophoblast there is a small cell cluster (embryoblast) denoting the nascent embryo.

By the close of the second week the embryoblast has become an embryonic disk. The disk is bilaminar, having a layer of ectoderm cells facing a small amniotic vesicle, and a layer of entoderm cells facing a yolk sac. It is shield-like in shape, being widest toward the end comprising the anlage of the head: about 65% of its total area becomes the head, and 35% the trunk and trunk derivatives. Its maximum linear dimension (distance from upper end of head region to lower end of trunk region) is about 0.008 inch, or 0.2 mm.

The embryo from two weeks to two months. Early in the third week a layer of mesoderm forms between the ectoderm and entoderm layers; this, in turn, becomes two layers each side of the organism's midline, with mesenchyme cells appearing between these layers. By the end of the week, beginnings of the circulatory, nervous, and musculo-skeletal systems are evident. These systems, represented by blood vessels, a neural groove, and somites respectively, each appear initially in the region of the upper trunk and future neck. The lateral portions of the embryonic disk have begun curving toward the front, that is, the organism is becoming cylindroid, with ectoderm tissue as it outer surface.

By the end of the first month, the embryo usually has a mouth cavity, gill arches and pouches, upper-limb buds, and an external tail. Commonly the heart has formed and become enclosed in the thorax, the stomach is beginning to form, and the pronephros (primitive kidney) is evident. The head is large relative to other parts of the embryo, and is bent strongly forward toward the thorax. There are individual differences: overall length of the embryo varies from 0.1 to 0.2 inch (less than 3 mm to more than 5 mm); the heart may be encased by the thorax, with no indication of limb buds; limb buds may be present, with the heart not completely enveloped and no sign of renal system rudiments.

During the first half of the second month, rudiments of the eyes, nose, lungs, and liver develop. Ossification (bone formation) begins in cartilages of the future collar and jaw bones (clavicles, mandible, and maxilla). The gill arches are transformed into bones of the inner ears, hyoid bone of the neck, and cartilages of the larynx. The gill pouches become eustachian tubes; thyroid, parathyroid, and thymus glands; and trachea.

By age 6 weeks the embryo has reached its maximum cephalocaudal (head-tail) curvature, the external tail has attained its greatest size, and the principal components of the limbs are indicated. The external tail rarely exceeds 0.05 inch (1.2 mm) in length, but this is sometimes equal to 15% of the embryo's total length. Lower-limb buds appear a few days after upper-limb buds, and in the first half of the second month the four limbs differentiate rapidly. By the end of the sixth week the arm, forearm, and anlage of manus (hand and wrist) are evident—as are the thigh, leg, and anlage of pes (ankle and foot). Overall length of the embryo in the middle of the second month averages about 0.5 inch, or 12 mm.

The typical embryo at age two months. Exclusive of umbilical cord, amniotic fluid, and encasing membranes, the embryo at age 2 months weighs slightly less than 0.1 ounce (about 2 gm). In the uncurved position, average body length is near 1.5 inches (39 mm); it consists of 44% head and neck length, 36% trunk length, and 20% lower limb length.

The head at age 2 months is large, the face is broad relative to its height, the eyes are widely spaced, the nose is broad, and the ears are low. Volume of the head and face is about 45% of total body volume. Head width (biparietal diameter) averages between 0.4 and 0.5 inch (near 11 mm), and is the greatest transverse dimension of the embryo. Interocular width (distance between the inner corners of the eyes) is slightly more than 50% of total face width, and width of the nose exceeds nose height by 45%. The ears are positioned at sites approximating the level of the lower jaw. Usually the central and two lateral sections of the upper lip are united by age 2 months, and union of the palate is completed soon after. Failures in union give cleft lip, cleft palate, or both. Clefts occur about once in 800 births; they are more common in males than females (about 60% to 40%), and more common for lip and palate jointly (55%) than for lip alone or palate alone.

Moving to the trunk, at age 2 months the thorax is circular, shoulder width narrow relative to head width, and hip width narrow relative to shoulder width. One month earlier the anteroposterior diameter (distance from front to back) of the thorax was almost twice the transverse diameter, now the two diameters are about equal, and—through continuing faster growth of thoracic width—at the middle of the prenatal period thoracic width surpasses depth by 18%. Shoulder width (distance between the most lateral points on the shoulder blades) is near 0.3 inch at age 2 months; comparatively, this average is 30% less than average head width and about equal to average lower limb length. Hip width is slightly under 0.2 inch (near 5 mm); this is about 60% of shoulder width and less than 45% of head width. Volume of the heart is about 20% of brain volume, and 4% of total body volume.

At age 2 months the limbs are small, there are individuated fingers and toes, the heels are formed, and there is beginning bone development in several segments of the limbs. Average length of each upper limb (0.4 inch) is less than head width or trunk length. Lower limb length (near 0.3 inch) is about 20% shorter than upper limb length, and more than 40% shorter than trunk length.

The feet no longer extend linearly below the legs; they have moved to positions almost at right angles with the legs. Distance from the heel to the end of the longest toe, on average, is between 0.1 and 0.2 inch—similar to total length of the embryo 1 month earlier. The middle toe projects farthest forward; for both hands and feet, the digital projection formula is 3>4>2>1>5.

Ossification has begun in several parts of the mesenchymal membrane of the cranial vault, and in cartilages of vertebrae, ribs, scapulae (shoulder blades), humerus and femur of the arm and thigh, radius and ulna of the forearm, tibia and fibula of the leg and, sometimes, distal segment of the thumb.

From the third month to mid-prenatal life. The human organism is designated an embryo between ages 2 weeks and 2 months, and a fetus during the remainder of prenatal life. In the third month the fetus shows differentiation of male and female genital organs; initiation of ossification in cartilages of the palms, soles, fingers, and toes; development of touch pads on the fingers; and beginning growth of nails on the fingers and toes. Previously unspecialized anlage of reproductive organs gives rise to either testes early in the month, or ovaries later in the month. In males, usually by the end of the month there are evidences of testes, penis, and scrotum.

Touch pads on the fingers are transitory; between the third and fifth months they slowly appear and disappear. Nail beds begin to form late in the third month: nails grow out gradually, reaching the ends of the fingers and toes about age 6 months.

Growth of hair begins in the fourth month. During this month fine and relatively long hair (lanugo) develops over much of the body surface. Coarser hair appears on the head in the fifth month, and as eyebrows and eyelashes in the seventh month.

Anlagen of the primary or lactal teeth (deciduous tooth buds) are identifiable as early as the middle of the second month. During the fourth and fifth months, enamel and dentin cells start building the crowns of these teeth; deposition begins about 14 weeks for primary central incisors, 15 weeks for primary first molars, 16 weeks for primary lateral incisors, 17 weeks for primary canines (cuspids), and 18 weeks for primary second molars.

At the midpoint of prenatal life (age 19 weeks) average body weight is about 14 ounces (400 gm); variations among individuals extend from 10.5 to 17.5 ounces. Average body length is 9.6 inches (24.5 cm). Averages are the same (3.1 inches) for lower limb length and length of the head and neck: it follows that total body length at this time is about 32% head and neck length, 36% trunk length, and 32% lower limb length.

Average head girth at age 19 weeks is 7.1 inches, exceeding chest girth by 1.2 inches, or 20%. Average shoulder width (2.2 inches) has become greater than head width (1.9 inches), but hip width

(1.6 inches) is still narrower than head width. Typically, upper limb length is 4 inches (10.2 cm); lower limb length is shorter by almost 1 inch, or about 23%. Other averages are 1.7 inches for calf girth, 1.6 inches for arm girth, and 1.4 inches for distance from the posterior surface of the heel to the tip of the longest toe. At age 19 weeks the longest toe is the first for some fetuses, and the second for others.

The human fetus between ages 19 and 38 weeks. In the fifth prenatal month, anlagen of the permanent incisor and first molar teeth are formed. Anlagen of the permanent premolars (these teeth later replace the primary molars) and the canines appear in later fetal months, and those of the permanent second and third molars after birth.

During the last 10 weeks of prenatal life there is slow migration of each male testis from the abdomen into the scrotum; this is brought about by gradual shortening (decrease in length by 75%) of Hunter's gubernaculum. Sometimes the testes do not complete their descent until after birth.

In both sexes, there is rapid increase in thickness of subcutaneous adipose tissue during the last couple of months prior to birth; this gives the skin a smooth, filled-out appearance contrasting with the wrinkled skin of a fetus born 2 to 3 months prematurely.

Dentin and enamel cells begin to form in the cusps of the permanent first molar teeth at differing times between 8 weeks before birth and 4 weeks after birth. During the last prenatal month, usually most of the lanugo is shed: not infrequently some remains at birth, particularly in the lower-back region.

Averages characterizing liveborn white fetuses at age 38 weeks are 19.9 inches (50.6 cm) for body length, 8.7 inches for upper limb length, 8.4 inches for trunk length, 6.6 inches for lower limb length, 5.0 inches for head and neck length, 4.5 inches for shoulder width and calf girth, 4.1 inches for arm girth, 3.7 inches for head width, 3.5 inches for hip width, and 7.3 pounds (3.3 kg) for body weight. Rarely does human body size at this age exceed 22.5 inches in total length, or 11 pounds in body weight.

Averages for body size and growth rate during fetal life. Presented in Table 1 are statistics for body length and body weight of white fetuses at eight prenatal ages. These statistics are derived from measures on more than 1,000 fetuses. Table 1 shows:

Table 1

Averages and Average Increments for Body Length and Body Weight of the Human White Fetus

Age in Weeks	Mean Size	Absolute Increase	Percentage Increase
Body Length (inches)			
10	3.2	----	----
14	6.1	2.9	91
18	8.9	2.8	46
22	11.5	2.6	29
26	13.9	2.4	21
30	16.0	2.1	15
34	18.0	2.0	13
38	19.9	1.9	11
Body Weight (pounds)			
10	0.04	----	----
14	0.27	0.23	575
18	0.7	0.43	159
22	1.4	0.7	100
26	2.5	1.1	79
30	3.9	1.4	56
34	5.5	1.6	41
38	7.3	1.8	33

1. The average of 19.9 inches for body length at age 38 weeks is more than six times that at age 10 weeks. The average of 5.5 pounds for body weight at age 34 weeks is more than twenty times that at age 14 weeks.

2. With advancing prenatal age, absolute increases (gains in inches and pounds) become smaller in average body length and larger in average body weight.

3. Throughout the fetal period, for both length and weight, there is slowing of the relative growth rate. Average body length increases about 90% between ages 10 and 14 weeks, near 30% between ages 18 and 22 weeks, 15% between ages 26 and 30 weeks, and slightly more than 10% between ages 34 and 38 weeks. Av-

erage increments in body weight for the same lunar months approximate 570%, 100%, 55%, and 33% respectively. Another way of expressing the declining growth rate is to note the amounts of time taken for body length to double and body weight to increase six times. Body length doubles in the 5-week period following age 10 weeks, and in the 16-week period following age 18 weeks. Body weight increases more than six times in the 4-week period following age 10 weeks, and less than six times in the 16-week period following age 22 weeks.

Growth is more rapid during the fetal period in hip width than head width, in trunk length than head and neck length, in lower limb length than trunk length, and in calf girth than arm girth. Increases in average size from age 10 weeks to age 38 weeks are near 400% for head and neck length, 500% for head width, 700% for trunk length, 770% for lower limb length, 880% for hip width, 1,500% for arm girth, and 2,100% for calf girth. Between ages 8 and 38 weeks, body length components change from 44% to 25% for head and neck length, and from 20% to 33% for lower limb length: these paired values register marked decline during prenatal life in predominance of the head and neck component, and substantial relative increase of the lower limb component.

Suggested Readings

Flanagan, G. L. **The first nine months of life.** New York: Simon and Schuster, 1962.

Heuser, C. H., and G. W. Corner **Contributions to Embryology, Carnegie Institution of Washington,** 1957, No. 244.

Jakobovits, A., and L. Iffy The rate of intrauterine growth in Caucasian, Black, and Central American populations between the 6th and 20th weeks of gestation, in O. G. Eiben (Ed.), **Growth and development, physique.** Budapest: Akadémiai Kiadó, 1977. (141-147)

Kraus, B. S., and R. Jordan **The human dentition before birth.** Philadelphia: Lea and Febiger, 1965.

*Meredith, H. V. **Child Development,** 1975, 46, 603-610.

Scammon, R. E., and L. A. Calkins **The development and growth of the external dimensions of the human body in the fetal period.** Minneapolis: University of Minnesota Press, 1929.

Schultz, A. H. **Quarterly Review of Biology,** 1926, 1, 465-521.

CHAPTER II

Body Size at Birth in Different Human Populations

Early studies. Considerable knowledge is available on the body size of white neonates born prior to 1890 in Australia, Europe, and the United States. On 120 Irish infants delivered in 1785 at a Dublin hospital, averages are 7 pounds (3.2 kg) for body weight, and 13.8 inches (35.1 cm) for head girth. From birth records amassed at Philadelphia during 1865-1872 on 712 United States infants, averages are 7.3 pounds for body weight, 19.2 inches for total length, and 12.6 inches for stem length (composite length of head, neck, and trunk; technically, distance from vertex to subischia plane). Averages for each sex separately, show males to exceed females by 8 ounces, 0.4 inch, and 0.2 inch in weight, total length, and stem length respectively.

On Australian infants delivered during 1857-1887 at Melbourne, averages for body weight are 7.3 pounds (3.3 kg) from 10,000 singleton neonates, and 5.5 pounds (2.5 kg) from 247 twin neonates. Singleton and twin averages for body length, obtained from measures on 400 infants born at Melbourne during 1861-1862, are 19.1 and 17.3 inches respectively. From data collected about 1875 on 630 German infants born at a hospital in Berlin, averages are 7.3 pounds (3.3 kg) for body weight, 19.7 inches (50.1 cm) for body length, 13.6 inches (34.6 cm) for head girth, and 3.5 inches (8.9 cm) for head width. Classification by birth order shows: on average, 291 first-born infants are slightly lighter, shorter, and smaller in head size than 339 later-born infants.

Recent findings on average body weight and length at birth. Statistics on groups of liveborn infants studied since 1960 are assembled in Tables 2 and 3. Table 2 pertains to offspring of parents residing at urban centers, and Table 3 to offspring of parents living in rural regions, or regions including both rural and urban areas. These tables, together with some complementary statistics, show:

1. Contemporary groups of live infants born in different parts of the world vary in average body weight from 5.1 pounds (2.3 kg) to 8.3 pounds (3.8 kg), and in average body length from 18.0 to 20.5 inches (46 to 52 cm). Averages are below 6 pounds and 18.5 inches

[9]

for Hindu* infants born at Calcutta and Vārānasi, and for Maya Amerinds* born in rural Guatemala. None of the averages on white infants born in Europe, New Zealand, or the United States are less than 7 pounds for weight, or 19.5 inches for length. Mean body weight of Swedish neonates at Malmö is 48% higher than that of Hindu neonates at Vārānasi: mean body length is 11% greater for Polish infants at Warsaw than for Hindu infants at Calcutta.

Table 2

Averages for Body Weight in Pounds and Body Length in Inches on Liveborn Infants of Both Sexes Measured Since 1960 at Urban Centers

Group	Number Measured	Mean Weight at Birth	Mean Length at Birth
Hindu, Vārānasi	510	5.1	18.4
Hindu, Calcutta	600	5.6	18.2
Hindu, Bombay and Delhi	2,703	6.3	19.0
Ceylonese, Kandy	813	6.4	------
Peruvian, Puno	2,676	6.7	------
Surinam Industani, urban	1,090	6.7	18.9
Shona black, Salisbury	5,919	6.9	------
Andean Quechua, Cuzco	619	6.9	------
Costa Rican, province capitals	150	6.9	19.5
English, Sheffield	708	7.0	------
Brazilian, Ribeirão Prêto	524	7.0	19.5
Japanese, urban	1,605	7.1	------
Italian, Bologna	200	7.1	19.9
Rhodesian white, Salisbury	2,492	7.2	------
Russian, Saratov	4,762	7.2	19.5
English, London	900	7.3	------
Peruvian, Lima	4,787	7.3	19.5
United States white, Charleston	1,995	7.4	------
Polish, Warsaw	1,297	7.4	20.0
New Zealand white, Christchurch	1,174	7.4	------
Swedish, Malmö	4,124	7.6	19.9
Quechua and Aymara, Tacna	1,920	8.0	------

* 'Hindu' is used to denote natives of India, and 'Amerind' as a contraction of American Indian.

Table 3

Averages for Body Weight in Pounds and Body Length in Inches on Liveborn Infants of Both Sexes Studied Since 1960 in Rural and Mixed Rural-Urban Areas

Group	Number Measured	Mean Weight at Birth	Mean Length at Birth
Lumi, New Guinea	63	5.3	------
Maya, rural Guatemala	456	5.6	18.0
Lunda, Angola	2,342	----	18.5
Mexican, Mexico	291	6.4	19.0
Ladino, Guatemala	47	6.6	18.9
Hindustani, Surinam	1,285	6.6	18.9
Bushnegro, Surinam	282	6.8	19.0
Mandinka, Gambia	100	6.9	------
Costa Rican, rural	200	6.9	20.0
Zuni, New Mexico	853	6.9	------
Puerto Rican, United States	1,848	7.0	------
Surinam Chinese	101	7.0	19.1
Uzbek, Uzbekistan	1,867	7.1	20.5
Navajo, Arizona and Utah	13,080	7.1	------
Surinam Creole	3,060	7.2	19.1
Kazakh, Kazakhstan	564	7.2	------
Arabian, Israel	41,245	7.3	------
Chilean, Chile	684	7.4	------
Tasmanian, Tasmania	1,114	7.4	20.3
Aleut, Aleutian Islands	486	7.4	------
Sioux, Northern Plains	4,388	7.5	------
Azerbaijani, Soviet Union	412	7.5	19.6
Cree, James Bay	768	8.3	------

2. The variation in means for birth weight of Amerind tribes extends from 5.6 pounds on Maya infants in Guatemala to 8.3 pounds on Cree infants in Canada. For groups in the United States, averages are 6.9 pounds on Hopi and Zuni; 7.1 pounds on Navajo and Pueblo; 7.3 pounds on Blackfeet, Eskimo, and Paiute; 7.5 pounds on Cherokee, Chippewa, Comanche, Creek, Crow, Kiowa, Pima, and Sioux; and 7.7 pounds on Cheyenne and Omaha. Means for Peruvian Amerinds are 6.9 pounds on infants born at Cuzco, ele-

vation 3.4 km, and 8.0 pounds on those born at Tacna, elevation 0.6 km. The Cree mean exceeds the means for Maya and Zuni by 48% and 20% respectively.

3. Compared with the average body length of liveborn Swedish neonates at Malmö, black neonates born in the northeast Lunda district of Angola are shorter by about 1.4 inches (3.5 cm). There is a similar difference between means on Polish infants at Warsaw and offspring of Bushnegro parents living in the interior of Surinam. For body weight, averages on contemporary black and white live, singleton infants born in the United States are near 6.8 pounds and 7.3 pounds respectively—a study in the mid-1960's obtained these values from delivery records collected in 11 states on 9,374 black and 8,698 white infants. The United States white average is similar to the averages in Table 2 for white infants at Charleston, Christchurch, London, Salisbury, Saratov, and Warsaw; and the United States black average similar to the averages in Tables 2 and 3 for black infants in Gambia, Rhodesia, and Surinam.

Average weight and length at birth on progeny of different socioeconomic classes. The largest differences between socioeconomic groups reported since 1960 are on Hindu infants at Bombay: compared with means on 270 offspring of parents representing "the best socioeconomic group," means on 1,000 progeny of parents living in "slum areas of the city" are smaller by 1.5 pounds in body weight and 2.0 inches in body length.

Differences in the same direction, but much smaller, are found in other urban populations. From a study on Hindu neonates at Delhi, averages are 0.5 pounds and 0.5 inches higher for 79 members of "professional and managerial" families than for 730 offspring of "unskilled" laborers. At Bello Horizonte, Brazil, averages are lower by 0.4 pounds and 0.3 inches on 1,000 infants born to parents of low to middle economic status than on 790 infants having parents with above average to high incomes. Samples consisting of 63 and 69 Colombian infants born to parents in the highest and lowest socioeconomic groups at Bogatá yield means differing by 0.5 pounds and 0.2 inches for weight and length respectively. At Charleston, South Carolina, means for body weight are 7.5 pounds on 1,480 white neonates with parents in the "above average" economic category, and 0.4 pounds less on 500 peers born into "below average" homes.

Differences among human populations in the percentage of live-

born infants weighing less than 5.5 pounds at birth. Often averages are reported in European and American studies using delivery records on infants weighing 5.5 pounds (2.5 kg) or more. This practice allows reasonably valid comparisons among white populations; it does not facilitate worldwide comparisons among infant populations.

Births of live infants weighing less than 5.5 pounds have been found to occur for 50% of 500 Hindu births at Vārānasi, 40% of 1,000 Hindu births at Delhi, 34% of 450 Maya births in rural Guatemala, 29% of 3,900 Burmese births at Rangoon to mothers unsupervised during pregnancy, 19% of 2,300 births at Rangoon to mothers who received pregnancy guidance and a daily supply of skimmed milk powder, 13% of 3,100 United States black births at Baltimore, 12% of 640 Guyana black births at Stoelmanseiland, 10% of 151,800 Hungarian births, 6% of 170,000 Jewish births in Israel, and 5% of 2,500 Finnish births at Helsinki.

It follows that average body weight is less for unselected liveborn infants than for infants who weigh 5.5 pounds or more at birth. In general, differences from the two types of sampling are largest for Hindu, Burmese, and Ceylonese infants; intermediate for infants of Chinese, Japanese, Filipino, and African black ancestry; and smallest for white infants of European ancestry. Means for body weight of unselected liveborn infants are less than corresponding means for infants weighing 5.5 pounds or more by amounts varying from 0.7 to 0.2 pounds.

Findings since 1960 on average head and chest girths of human neonates. Averages at birth from contemporary studies of head girth and chest girth are displayed in Table 4. In other neonatal studies since 1960 measures of head girth, but not chest girth, were taken: means are 12.7 inches (near 33 cm) on 600 infants at Calcutta; 13.2 inches on 813 infants at Kandy; 13.3 inches on 273 Jewish infants at Jerusalem; 13.4 inches on 331 English infants at London, also on 525 Sudanese infants at Khartoum and Omdurman; 13.6 inches on 4,124 Swedish infants at Malmö; and 13.8 inches (35 cm) on 107 Iranian infants at Shiraz born to mothers "of the highest socioeconomic class," receiving superior prenatal supervision. From a Ceylonese study at Colombo, means are 11.9 inches (near 30 cm) on 159 newborn infants weighing 5.0 pounds or less, and 13.4 inches (34 cm) on 150 infants weighing 5.5 pounds or more.

Findings for head and chest girths are as follows:

Table 4

Averages for Head and Chest Girths in Inches on Groups of Newborn Infants of Both Sexes Measured Since 1960

Group	Number Measured	Mean Head Girth	Mean Chest Girth
Liveborn Infants Regardless of Birth Weight			
Hindu, Vārānasi	501	12.5	11.5
Maya, rural Guatemala	456	12.6	11.8
Hindu, Agra and Kampur	1,500	12.9	11.9
Uganda black, Kampala region	402	13.0	12.5
Ladino, rural Guatemala	40	13.1	12.1
United States black, New Orleans	200	13.3	12.4
English, Sheffield	106	13.5	12.2
Peruvian, Lima	4,787	13.5	13.2
Brazilian, Ribeirão Prêto	524	13.6	13.0
Italian, Bologna	200	13.8	13.3
Viable Infants Weighing 4.4 Pounds or More			
Colombian, Medellin*	1,650	13.2	12.8
Hindu, Delhi	1,725	13.4	12.7
Sardinian, Sassari province**	200	13.5	12.8
Quechua, Cuzco and Lima**	153	13.5	13.0
Polish, Lublin	290	13.6	13.0
Bulgarian, Sofia region**	300	13.7	13.0
Italian, Friuli province**	500	13.7	13.2
Cambodian, Pnompenh area**	200	13.7	13.6
Azerbaijani, Soviet Union**	412	13.9	13.5
Russian, Volgograd**	2,000	14.0	13.6

* Birth weigh between 4.4 and 8.8 pounds.
** Birth weight 5.5 pounds and up.

1. Head girth of different human populations at birth, on average, varies from 12.5 to 14.0 inches. Averages are less than 13.0 inches for liveborn Maya neonates of rural Guatemala, and Hindu neonates at Agra, Calcutta, Kampur, and Vārānasi. On Azerbaijani, Brazilian, Cambodian, and Iranian groups, also white groups at Bologna, Lublin, Malmö, and Volgograd, averages are greater than 13.5 inches.

2. Those groups in Table 4 that are below 13 inches in mean head girth are below 12 inches in mean chest girth. Means for chest girth exceeding 13 inches are found for Cambodians at Pnompenh, Italians at Bologna, Peruvians at Lima, Russians at Volgograd, and Azerbaijanis in the rural district of Divichinsk. The mean for Italian neonates at Bologna is 15% higher than that for Hindu neonates at Vārānasi.

3. At birth, average head girth is larger than average chest girth. This is shown consistently in Table 4 on samples of newborn infants studied in Africa, Asia, Europe, and North and South America. For Hindu groups, mean chest girth is 92% to 93% of mean head girth; for Bulgarian, Italian, Polish, and Russian groups, corresponding values are between 95% and 97%.

Stem length, lower limb length, and the skelic index at birth. Averages for stem length (vertex-subischia distance) are accessible on two groups studied in the mid-1960's—Hindu and Quechua neonates—and two groups studied around 1950—United States black and white neonates. In each instance, measures were taken on newborn infants weighing at least 5.5 pounds: means are 12.8 inches (32.5 cm) for 368 Hindu infants at Agra, 12.9 inches for 153 Quechua infants at Cuzco and Lima, 13.2 inches for 272 American black infants at Philadelphia, and 13.4 inches (34 cm) for 738 American white infants at Philadelphia. The entire group measured at Agra included 132 Hindu live infants weighing less than 5.5 pounds: the mean on pooling all 500 measures is 12.4 inches (31.5 cm).

Lower limb length is conveniently obtained as total length (vertex-soles distance) minus stem length. At Agra, the mean for Hindu liveborn infants is 6.1 inches, and that on the subgroup weighing at or above 5.5 pounds is 6.4 inches. Means on other groups of infants weighing no less than 5.5 pounds are 6.3, 6.5, and 6.5 inches for Pennsylvanian white, Pennsylvanian black, and Peruvian Amerind neonates respectively.

Average lower limb length of infants at birth is about equal to one-half average stem length. Specifically, lower limb length relative to stem length is typically 47% on Philadelphia white infants, 49% on Philadelphia black infants, and 50% on Agra Hindu and urban Quechua infants. From other recent studies, average skelic indices (lower limb length x 100/stem length) are 46 on Jewish infants at Jerusalem, 51 on Jamaican infants at Kingston, and 54 on Hindu infants at Delhi. Generalizing, for different populations of newborn infants average lower limb length falls between 46% and 54% of average stem length.

Sex differences in body size of human populations at birth. Average body weight at birth is slightly less for females than males. From samples of 5,000 or more individuals of each sex, males are found to exceed females by 2 ounces for Hindu infants, 3 ounces for Japanese and Chinese infants, 4 ounces for United States white and Amerind infants, and 5 ounces for Polish and Senegal black infants.

Typically, averages for body length are less for females than males by 1.0% to 1.5%; means from large samples usually show absolute differences between 0.2 and 0.3 inch. Sex differences for head girth and chest girth are similar in direction and size to those for body length.

Individual differences in body size at birth. For body weight, individual differences within and between human groups can be illustrated by simultaneously describing and comparing records on 21,000 Hindu live births at Bombay, and 7,200 English live births in Middlesex County. Birth weights are under 3 pounds for 2.2% and 0.4% of the Hindu and English infants respectively. Keeping the same order, 4.1% and 0.7% of birth weights are between 3 and 4 pounds, 82% and 28% from 4 to 7 pounds, 10% and 38% from 7 to 8 pounds, 1.7% and 24% between 8 and 9 pounds. Birth weights above 9 pounds occur for none of the Bombay infants, and for 8.9% of the Middlesex infants. Wider differences exist between the Hindu infants born at Bombay and Cree infants born in the James Bay area of Canada: of 768 Cree infants, 0.4% weigh less than 4 pounds, 15% between 4 and 7 pounds, 30% between 8 and 9 pounds, and 29% above 9 pounds.

Findings in the metric system on 985,000 Polish liveborn neonates show variations from 0.2% below 1 kg (2.2 pounds) to 0.2% above 5 kg (11 pounds): intermediate relative frequencies are 2.6%, 21%, 66%, and 10% for neonates with weights from 1 to 2 kg, 2 to 3 kg, 3 to 4 kg, and 4 to 5 kg respectively

Measures of body length on 2,300 liveborn progeny of black parents residing in northeast Angola show 50% are either shorter than 17.6 inches, or longer than 19.4 inches. The values beyond which 50% scatter (conversely, between which 50% cluster) are 18.2 and 20.1 inches for 800 Creole neonates of rural Surinam, 18.4

and 20.2 inches for 880 Papuan neonates of Sorong and Biak Island, 19.1 and 20.6 inches for 1,600 Norwegian neonates at Bergen.

Information on individual differences in several measures of body size is provided in a study on over 700 United States white children born at Philadelphia, and weighing 5.5 pounds or more at birth. One-fourth of each series of measures lies below 6.8 pounds (body weight), 19.1 inches (body length), 12.9 inches (stem length), 13.1 inches (head girth), 12.3 inches (chest girth), 2.9 inches (hip width between the crests of the ilia), and 4.2 inches (calf girth). Another one-fourth lies above 8.1 pounds for body weight, 20.2 inches for body length, 13.7 inches for stem length, 13.8 inches for head girth, 13.2 inches for chest girth, 3.3 inches for biiliocristal hip width, and 4.8 inches for calf girth. Standard deviations are near 0.9 pounds for body weight, 0.8 inch for body length, 0.6 inch for stem length and chest girth, 0.5 inch for head girth, 0.4 inch for lower limb length and calf girth, and 0.2 inch for hip width.

Suggested Readings

Adams, M. S., and J. D. Niswander **Human Biology,** 1973, 45, 351-357.

Bakwin, H., and R. M. Bakwin **Human Biology,** 1934, 6, 612-626.

Chrzastek-Spruch, H. **Prace I Materialy Naukowe, Instytut Matki I Dyiecha,** 1968 ,11, 65-104.

*Kasius, R. V., A. Randall, W. T. Tompkins and D. G. Wiehl **Milbank Memorial Fund Quarterly,** 1958, 34, 335-362.

Meredith, H. V. **Human Biology,** 1952, 24, 290-308.

*Meredith, H. V. Growth in body size: a compendium of findings on contemporary children living in different parts of the world, in H. W. Reese (Ed.), **Advances in child development and behavior.** New York: Academic Press, 1971 (Volume 6, 153-238).

Singer, B., L. Blake and J. Wolfsdorf **South African Medical Journal,** 1973, 47, 2399-2402.

Chapter III

Body Size in Relation to Birth Order

Body size of first-born and later-born neonates. This chapter pertains to the body size of progeny from successive pregnancies. To date, the largest number of studies on the topic compares average birth weight of first-born infants with that of siblings born later. Table 5 presents statistics for several populations on the difference in average body weight at birth of first-born and later-born infants.

In some instances, data for the construction of Table 5 were gathered at two or more identified locations. These locations are as follows:

United States black—at Baltimore, Nashville, and urban hospitals in Louisiana, Massachusetts, New York, Pennsylvania, and Virginia.

Italian—at Bologna, Cagliari, Genoa, Perugia, and Pisa.

British—at Edinburgh and London, and in Middlesex County.

German—at Berlin, Bonn, Dresden, Erlangen, Greifswald, Halle, Jena, Königsberg, Leipzig, Marburg, Munich, Strassburg, Wuppertal, and Würzburg.

Chinese—at Hong Kong, Kowloon, Peking, and Singapore.

Hindu—at Bombay, Calcutta, Dabra, and Durban.

Norwegian—at Bergen and Oslo.

United States white—at Baltimore, Boston, Iowa City, Los Angeles, Minneapolis, New York, Philadelphia, and urban hospitals in Connecticut, Louisiana, Oregon, Rhode Island, Tennessee, and Virginia.

African black—at Dakar and Durban.

Table 5 supports two generalizations:

1. It is a worldwide finding that first-born human offspring, on average, weigh less at birth than those of all other birth ranks combined.

2. First-born infants of different populations average less in body weight than later-born infants by amounts between 4 and 9 ounces. Typically, first-born offspring are slightly more than 6 ounces (180 gm) lighter than those pooled from later pregnancies.

Table 5

Differences in Ounces Between Average Birth Weight of First-born Infants and Average Birth Weight of Infants from Subsequent Pregnancies

Group	Number Weighed First-born	Number Weighed Later-born	Later-born Mean Higher than First-born Mean by:
Israeli Jewish	51,886	119,532	4.6
Hungarian	59,728	61,516	4.7
United States black	4,586	10,047	5.0
Italian	3,562	4,016	5.1
Tasmanian	462	649	5.3
British	9,199	9,080	5.4
German	14,565	15,913	5.5
Chinese	6,478	18,848	5.5
Austrian, Vienna	2,590	2,252	5.6
South African white	1,551	1,605	5.6
Hindu	6,694	20,747	6.3
Norwegian	7,184	7,076	6.7
United States white	16,670	28,330	6.8
French, Paris	1,620	2,269	6.9
African black	2,453	8,125	7.4
Israeli Arabian	6,351	34,123	8.1
Philippine Islander	457	780	8.1
Danish, Nykobing district	1,283	4,540	8.5
Papuan, Sorong	87	266	8.8

Average body length at birth is less for first-born neonates than for the aggregate of neonates with other birth ranks by about 0.14 inch (3.5 mm). For Filipino, Italian, German, Hungarian, and United States white groups, first-born neonates are shorter than those later in family series by average amounts between 0.1 and 0.2 inch. Sample size of the first-born subgroups falls between 450 and 11,500, and sample size of the later-born subgroups varies from 780 to 13,100. Smaller samples of Mexican, Norwegian, and United States black neonates give differences in the same direction, and of similar size. Among 560 white, singleton, viable infants born at Baltimore, first-born offspring comprised 60% of those measuring less than 19.7 inches (50 cm).

For several measures of body size—including head girth, head width, face width, stem length, chest girth, shoulder width, and hip width—means obtained in a New York study of white progeny are slightly smaller from 810 first-born neonates than from 840 later-born neonates. Analyses of other data support these findings for head size, and extend them to lengths of the upper and lower limbs.

Size at birth from more than twofold grouping on birth order. For several of the human populations represented in Table 5, findings are accessible on average birth weight of no less than three birth-order subgroups. The three subgroups studied most frequently are: first-born, second- through fourth-born, and all birth ranks above fourth-born. A less commonly formed set of birth-order categories is 1, 2-3, and 4 or higher. Table 6 is derived from such trichotamous schemata. This table shows that typically first-born infants weigh about 8.5 ounces (240 gm) less than those at ordinal positions later than fourth-born.

Taken together, Tables 5 and 6 reveal:

1. Body weight at birth, on average, increases with birth order from the first offspring to beyond the fourth in family series.

2. Among birth-order subgroups of specific populations, average body weight of first-born neonates is surpassed less by average weight of 'second and up' neonates (Table 5) than by that of 'fifth and up' neonates (Table 6).

Additional studies in Europe and the United States have:

1. Compared successive progeny of the same parents and found average weight at birth is less for the first child than the second, and less for the second child than the third.

2. Made analyses for each birth rank, demonstrating the difference in average birth weight between first- and second-born infants is greater than between second- and third-born infants, and subsequent differences become gradually smaller.

A study of European infants comparing means for body length finds 8,610 first-born neonates to be shorter than 1,620 fifth- and later-born neonates by 0.2 inch (0.5 cm). Using an alternative method of comparison, a study on several hundred United States white infants finds body lengths shorter than 19.7 inches (50 cm) occur more frequently among first-born offspring than among fifth-born offspring.

Table 6

Differences in Ounces Between Average Birth Weight of First-born Infants and Average Birth Weight of Infants Born Fifth or Later in Family Series

Group	Number Weighed First-born	Fifth and Up	Mean for 'Fifth and Up' Exceeds First-born Mean by:
Hungarian	59,728	8,274	5.6
English, London	1,547	562*	6.3
Israeli Jewish	51,886	27,851	7.0
United States black	3,862	3,012	7.0
Chinese	6,478	6,819	7.3
English, Middlesex County	420	162	7.4
Italian, Perugia	378	294	7.7
Hindu	6,694	8,095	8.3
German	9,685	1,488	8.4
Austrian	2,202	312	8.5
Italian, Bologna	133	237*	8.8
United States white	14,776	8,416	9.1
German, Halle	943	205*	9.2
African black	2,453	2,726	9.6
Israeli Arabian	6,351	19,141	9.9
Danish, Nykobing district	1,283	1,821	10.2

* Fourth and higher birth ranks.

Birth order and body size in late infancy and at older ages. Investigations on 2,700 German infants at Halle and Frankfurt-am-Main, and 1,200 Scottish infants at Glasgow, show that the low positive relationship between birth order and body weight at birth becomes reversed to a low negative relationship by the close of the first postnatal year. Longitudinal data on 180 Polish individuals at Warsaw support and extend this finding by revealing that growth velocities for weight and length during infancy are higher on first-born infants than those from third and later pregnancies in parental series.

On Tasmanians age 3 years, averages for body weight are 33.5 pounds for 462 first-born children, and 32.8 pounds for 314 children with family ranks beyond second-born. At age 5 years, averages for standing height on United States white children at Iowa City are 43.6 inches and 43.3 inches on 120 children each in the first-born and 2-4 birth-order categories. At age 6 years, height and weight averages for English children at Middlesbrough are higher on 200 first-born than 600 later-born individuals by 1.0 inch and 1.5 pounds. For Spanish boys between ages 6 and 10 years residing in high-income homes at Madrid, averages on 990 first-born boys are higher than those on 490 boys born subsequent to the fourth child by 0.6 inch and 2.1 pounds.

At ages between 7 and 10 years, height and weight averages for inhabitants of Aruba Island are higher on 260 children of birth orders 1 and 2 than on 135 peers of birth orders "5 and up" by 1.1 inches and about 5 pounds. At ages between 7 and 15 years, there are differences in the same direction on Canadian children and youths at Toronto, and United States white children and youths at Oakland. At ages 15, 16, and 17 years, 595 first-born Spanish male youths at Madrid are taller and heavier than 295 comparable youths born at sibling positions "5 and up" by 1.1 inches in height and 7.9 pounds in weight.

Averages for height of Norwegians age 20 years measured in the early 1960's decrease from 70.1 inches on 2,700 first-born men, through 69.7 inches on 730 third-born men, to 69.4 inches on 510 men in the birth-order category "5 and up": the first-born subgroup is 1% taller than the "5 and up" subgroup.

Suggested Readings

Barron, S. L., and M. P. Vessey **British Journal of Preventive and Social Medicine,** 1966, 20, 127-134.

Bernis, C. Some aspects of growth in Spanish children, in O. G. Eiben (Ed.), **Growth and development, physique.** Budapest: Akadémiai Kiadó, 1977. (279-290)

Chrzastek-Spruch, H. Some genetic and environmental problems of physical growth and development of children aged 0-7 years, in O. G. Eiben (Ed.), **Growth and development, physique.** Budapest: Akadémiai Kiadó, 1977. (35-42)

Coy, J. F., I. C. Lewis, C. H. Mair, E. A. Longmore and D. A. Ratkowsky **Medical Journal of Australia,** 1973, 2, 12-18.

*Grossman, S., Y. Handlesman and A. M. Davies **Journal of Biosocial Science,** 1974, 6, 43-58.

*Meredith, H. V. American Journal of Physical Anthropology, 1950, 8, 195-224.

*Warburton, D., and A. F. Naylor American Journal of Human Genetics, 1971, 23, 41-54.

CHAPTER IV

Tobacco Smoking in Pregnancy and Body of Progeny

In relation to the topic of the preceding chapter, the topic of this chapter has a much shorter research history. Several human studies on the association between body size and birth order were reported during the 1860's; human studies on the association between body size and maternal tobacco smoking during pregnancy were first published a little before 1960.

Average difference in birth weight for progeny of non-smoking and tobacco smoking mothers. Measures of infant body weight at birth, and records on whether the mother used tobacco during pregnancy have been obtained on human populations in France, Finland, Great Britain, Italy, New Zealand, Sweden, and the United States. Listed in Table 7 are statistics showing the difference between means for birth weight on offspring of non-smoking mothers and offspring of mothers who, during pregnancy, smoked cigarettes made from tobacco.

Comment relating to seven rows of Table 7 follows:

French—the smoking mothers are from a "a higher social class" than the non-smoking mothers.

Finnish—records showing pregnancy "complicated by Rh-immunisation or toxemia" are excluded.

New Zealand white—these Christchurch samples include stillborn infants weighing 2.2 pounds (1 kg) or more.

United States black—records are combined from California, Maryland, Massachusetts, New Jersey, and South Carolina; they largely pertain to families of middle to low socioeconomic status.

United States mixed—these data were gathered at Seattle, Honolulu, and United States "worldwide naval installations"; 88% of the subjects are white and 12% black, Oriental, or Polynesian.

British—materials are pooled from Aberdeen, Birmingham, Dublin, Sheffield, and other parts of the British Isles.

United States white—these records were compiled in California,

[25]

Colorado, Kansas, Maryland, Massachusetts, Nebraska, New Jersey, and South Carolina; the parents were predominantly urban residents.

Table 7

Difference in Ounces Between Means for Body Weight at Birth on Infants Whose Mothers Did or Did Not Use Tobacco During Pregnancy

Group	Subgroup Size Non-smoking	Smoking	Progeny of Smoking Mothers Lighter by:
French, Paris*	5,500	1,100	4.2
Finnish, Helsinki*	1,827	741	4.2
Swedish, Malmö and Uppsala	5,033	2,581	5.5
New Zealand white	717	457	5.6
United States black	9,300	6,560	6.0
Italian, Ferrara	1,145	430	6.0
United States mixed*	26,200	25,300	6.1
British	21,100	9,800	6.2
United States white	15,160	13,200	7.8

* Singleton neonates only.

Table 7 supplies a foundation for two generalizations:

1. Average body weight at birth is less for progeny of mothers who smoke tobacco during pregnancy than for progeny of mothers not using tobacco. This is found consistently.

2. The mean birth weight for offspring of women who engage in tobacco smoking during pregnancy is lower than that for offspring of women not using tobacco at this time by an average amount near 6 ounces.

Average amount by which birth weight is higher for infants of non-smoking mothers than for infants of mothers who, in pregnancy, smoke 10 cigarettes or more daily. Table 8 is similar to Table 7 except that fewer groups are represented, and the comparisons are for body weight of newborn infants from pregnancies with no maternal tobacco smoking, and with moderately heavy (10 or more cigarettes daily) maternal tobacco smoking.

From Table 8, it is found: Mean body weight at birth is about 7.5 ounces (210 gm) higher for offspring of women who do not use tobacco during pregnancy than for offspring of women who smoke 10 or more cigarettes daily.

A study on United States white, live, singleton infants born in Massachusetts obtained information on maternal tobacco smoking during pregnancy using four categories for number of cigarettes smoked daily: none, less than 10, 11 to 20, more than 20. The number of mother-child pairs in each category is large—5,960, 1,250, 2,520, and 2,370. Mean birth weight on the infants of non-smoking mothers is higher than the means for succeeding categories by 3 ounces, 8 ounces, and 10 ounces respectively. Findings from several other studies show similar birth weight reduction out to the "20 or more" category. In short, there are substantial grounds for concluding that as maternal use of tobacco during pregnancy increases, birth weight of progeny decreases.

Table 8

Difference in Ounces Between Means for Body Weight at Birth on Issue of Mothers Who, in Pregnancy, did not use Tobacco or Smoked 10 or more Cigarettes Daily

Group	Cigarette Subgroup		Progeny of Non-smoking Mothers Heavier by:
	None	10 or more	
Swedish, Malmö and Uppsala	5,033	527	6.6
United States black	3,791	819	6.7
United States mixed*	26,200	17,050	7.2
British	18,060	4,125	7.8
Italian, Ferrara	1,145	203	7.8
New Zealand white	717	208	7.8
United States white	7,740	6,090	9.5

* Primarily of white ancestry, a small percentage had Afro-black, Chinese, Japanese, or Polynesian progenitors.

Frequency with which body weight of newborn infants does not exceed 5.5 pounds in three categories of maternal tobacco smoking during pregnancy. Table 9 brings together statistics from European and North American studies on the relative frequency of birth weights at or below 5.5 pounds from pregnancies during which mothers (1) did not use tobacco, (2) smoked undetermined amounts of tobacco, or (3) smoked 10 or more cigarettes daily.

The smoking relationship exhibited in the upper section of Table 9 is supported by additional reports on United States black and white neonates, and by reports of smaller scope on British, German, and New Zealand neonates. A British study compared two groups of singleton offspring, one including 225 unselected pregnant women, and the other 90 women whose progeny weighed 5.5 pounds or less. Relative frequencies for mothers smoking cigarettes during pregnancy are 36% in the unselected group, and 50% in the low birth weight group.

Examination of Table 9 shows:

1. The percentage of infants weighing no more than 5.5 pounds at birth increases with the quantity of tobacco mothers use during pregnancy. Citing the largest sampling in the table (United States mixed), birth weights not exceeding 5.5 pounds are found for about 6% of progeny of non-smoking mothers, 9% of progeny of mothers smoking at all in pregnancy, and 10% of progency of mothers who smoked 10 or more cigarettes daily.

2. Besides the systematic finding for separate populations in frequency of low birth weight across maternal smoking categories, there are differences between ethnic groups. Birth weights at or below 5.5 pounds are found at relative frequencies near 3.5%, 6.5%, and 7.5% for United States white infants, and near 9%, 15%, and 18% for United States black infants.

Maternal tobacco smoking during pregnancy in relation to frequency of newborn body weight above 5.5 pounds. Among the mothers of 3,900 Swedish infants having birth weights greater than 5.5 pounds, 69% did not use tobacco during pregnancy, and 6% smoked 10 or more cigarettes daily. Corresponding values are 64% and 10% among 5,580 United States white neonates, and 51% and 26% among 197 United States black neonates.

In a study of 2,000 British singleton infants, those with birth weights at or above 8 pounds were selected for separate analysis. Of these, 78% were offspring of non-smoking mothers, and 22% offspring of mothers who smoked tobacco during pregnancy.

Maternal tobacco smoking during pregnancy in relation to birth length of progency and neonatal mortality. The influence of maternal tobacco smoking on average body length of infants at birth has been investigated in Finland, France, Sweden, and the United States. Means for offspring of mothers who smoke during pregnancy are lower than those for offspring of non-smoking mothers by about 0.2 inch on 1,800 Finnish and 3,500 French neonates, 0.3 inch on 2,800 Swedish neonates, and 0.4 inch on 785 United States neonates. When pooled, these studies yield a difference near 0.2 inch (about 5 mm) as the amount by which birth length for infants of non-smoking mothers exceeds that for infants of cigarette-smoking mothers.

A study on the progency of Irish women at Dublin finds deaths during the first postnatal month are 12 per 1,000 for 6,400 infants of non-smoking mothers, and 18 per 1,000 for 4,900 infants of

Table 9

Percentage of Newborn Infants Weighing 5.5 Pounds or Less from Pregnancies of Maternal Non-smokers and Two Classes of Tobacco Smokers

Group	Subgroup Size		%-age At or Under 5.5 Pounds	
	Non-smoking	Smoking	Non-smoking	Smoking
Offspring of Non-smoking Mothers Compared with Those of Mothers Smoking*				
Swedish, Malmö	2,820	1,300	3.3	5.0
Finnish, Helsinki	1,827	741	4.3	5.7
United States white	10,450	8,570	3.6	6.7
French, Paris	5,500	1,100	4.0	8.1
Scottish, Aberdeen	867	832	4.2	8.7
United States mixed**	31,000	26,900	5.8	9.1
United States black	5,460	2,980	9.0	14.9
Progeny of Mothers in Cigarette-smoking Classes 'None' and '10 or More'				
Swedish, Malmö	2,820	270	3.3	5.6
United States white	10,450	4,660	3.6	7.6
United States mixed**	31,000	17,600	5.8	9.9
United States black	5,460	980	9.0	17.8

* In this section of Table 9, 'smoking mothers' denotes: Mothers who smoked during pregnancy, disregarding the quantity of tobacco used.
** Predominantly of white ancestry, small numbers with Afro-black, Chinese, Japanese, Mexican, and Polynesian ancestry.

mothers smoking 5 or more cigarettes daily throughout pregnancy. Corresponding values are 15 and 21 per 1,000 from 12,000 United States white mother-child records and, from 4,700 United States black mother-child records, 16 and 22 per 1,000.

A few studies on maternal use of tobacco in pregnancy do not find mortality to be higher for issue of smoking than non-smoking mothers. However, a report evaluating the accumulated findings on perinatal and neonatal mortality from 27 samples concludes: In the aggregate, research indicates maternal tobacco smoking is associated with higher infant mortality.

Maternal tobacco smoking during pregnancy in relation to postnatal body size. A study on United States white, singleton infants reports identical means for body weight at the end of the first postnatal year on 2,600 infants born to non-smoking women, and 2,050 infants born to women who, during pregnancy, smoked 10 or more cigarettes daily. Other studies in Great Britain and the United States find small differences at this age. On the whole, at age 1 year following birth offspring of mothers using and not using tobacco during pregnancy either do not differ in average body weight, or differ by no more than 0.7 pounds (300 gm).

Research on British and North American children indicates that maternal tobacco smoking during pregnancy has a slightly lowering effect on average standing height in childhood. From rigorously conducted studies involving more than 15,000 children, it is found that progeny of mothers who do not use tobacco during pregnancy are taller than those of mothers smoking 10 or more cigarettes daily by about 0.3 inch at ages 4 and 5 years, 0.4 inch at age 7 years, and 0.5 inch at age 11 years.

Suggested Readings

Butler, N. R., and H. Goldstein **British Medical Journal,** 1973, 4, 573-575.

Kullander, S., and B. Källén **Acta Obstetricia et Gynecologica Scandinavica,** 1971, 50, 83-94.

MacMahon, B., M. Alpert and E. J. Salber **American Journal of** Epidemiology, 1965, 82, 247-261.

*Meredith, H. V. **Human Biology,** 1975, 47, 451-472.

Penchaszadeh, V. B., J. B. Hardy, E. D. Mellits, B. H. Cohen and V. A. McKusick **Johns Hopkins Medical Journal,** 1972, 131, 11-23.

Rush, D., and E. H. Kass **American Journal of Epidemiology,** 1972, 96, 183-196.

CHAPTER V

Body Growth During Human Infancy

Loss and gain in body weight during the first ten days of postnatal life. By 1870 it had been found that average body weight of human infants is less 1 week after birth than at birth. This finding prompted several questions: Does physiological loss occur, or is the decrease due solely to discard of extraneous material from the digestive tract and outer body surface? If physiological loss occurs, how many days after birth does it continue? Can the duration and amount of loss be reduced by extra-maternal feeding procedures?

In the mid-1870's an investigator at Copenhagen pioneered in discovering that physiological weight loss (1) almost always occurs, (2) typically continues for about 3 days, and (3) can partly be reduced by early feeding procedures. Later investigators verified and augmented these outcomes.

Nutritional substances fed prior to the availability of maternal breast milk include milk of a wet-nurse; a weak solution of gelatin, milk, and sodium chloride; water, milk, and dextri-maltose; and lactic acid milk with corn syrup. Of the experimental diets tried, few reduced neonatal weight loss more than 50%

One of the major studies on change in body weight during the first 10 days after birth is based on 1,000 physically normal infants weighing 5.5 pounds or more at birth, receiving water at 4-hour intervals from the hour of birth, and fed at the breasts of their mothers. Both parents were of northwest European ancestry, and the mothers were healthy throughout pregnancy. Each child was washed in oil before birth weight was taken, then weighed daily on 9 successive days. The study finds:

1. There is loss in mean body weight during the first 3 postnatal days of 7.6 ounces (215 gm), and gain during the next 4 days of 4.2 ounces (120 gm). In consequence, mean body weight is higher at birth than 1 week later by 3.4 ounces.

2. Within the group, less than 2% show no neonatal weight loss, and 2% lose weight for 6 to 8 days. Duration of loss is 1 day for 6%, 2 to 4 days for 84%, and 5 days for 6%.

3. The average amount of loss in weight, determined from birth

[31]

weight and weight on the day it is lowest, approximates 8.7 ounces (250 gm). There is no difference in average loss for males and females, but average loss increases slightly with birth order (from 8.3 ounces for first-born infants to 9.5 ounces for fifth-born infants). Individual differences vary from no weight loss to a loss of 1.3 pounds.

4. For about 1 infant in 4, body weight 1 week after birth equals or surpasses birth weight. At age 10 days, 50% of infants weigh as much or more than weight at birth. Compared with infants weighing between 5.5 and 6.5 pounds at birth, those weighing over 8.5 pounds have greater weight loss, and are later in attaining a weight equivalent to birth weight.

Increase in body weight during each half of the first postnatal year. Averages for amount of gain in body size during the first and last halves of the year following birth are displayed in Table 10. On three of the groups represented, data were collected at the following places:

United States black—at Detroit, New Haven, New Orleans, New York, Philadelphia, and Washington, D. C. for weight and length. At New Orleans, Philadelphia, and Washington, D. C. for head girth; Philadelphia and Washington, D. C. for chest girth; and Philadelphia for hip width.

United States white—at Berkeley, Boston, Chicago, Cleveland, Denver, Detroit, Iowa City, Minneapolis, Newark, New Haven, New York, Philadelphia, and Yellow Springs for weight and length. At Berkeley, Boston, Denver, Iowa City, Newark, New Haven, and Philadelphia for head girth; Boston, Denver, Newark, and Philadelphia for chest girth; and Boston, Denver, and Philadelphia for hip width.

United States twins—at and near Louisville, Kentucky, on twin offspring of white parents.

Table 10 shows:

1. Body growth is much more rapid during the first 6 months after birth than during the succeeding 6 months. For United States white singleton infants, average gains from birth to the middle of the first year are near 10 pounds in body weight, 6.5 inches in body length, 3.5 inches in head girth, 4.5 inches in chest girth, and 1.5 inches in hip width. Between ages 6 and 12 months average increases are about half as large for weight and length, and less than half as large for head girth, chest girth, and hip width.

Table 10

Average Increments During Semiannual Periods of Infancy in Body Weight, Body Length, Head Girth, Chest Girth, and Hip Width

Group	Birth to 6 Months		6 to 12 Months	
	Number	Mean Gain	Number	Mean Gain
Body Weight in Pounds				
Hindu, Delhi	1,157	7.2	973	3.6
French, Paris	448	8.8	406	4.8
United States black	2,380	9.4	1,560	5.2
United States white	3,020	9.9	3,030	5.0
United States twins	583	10.3	546	5.0
Body (Vertex-soles) Length in Inches				
Hindu, Delhi	1,157	5.6	973	2.8
French, Paris	448	6.2	406	3.1
United States white	3,385	6.5	3,185	3.4
United States black	944	6.7	1,025	3.3
United States twins	172	6.8	548	3.5
Head Girth in Inches				
Hindu, Delhi	1,157	2.9	973	1.0
United States white	2,094	3.4	1,828	1.2
Polish, Lublin	252	3.5	248	1.2
United States black	527	4.0	443	1.2
United States twins	172	3.9	538	1.3
Chest Girth in Inches				
Hindu, Delhi	1,157	3.2	973	1.1
United States black	307	4.4	228	1.3
United States white	937	4.4	680	1.4
Polish, Lublin	252	4.3	248	1.5
Hip (Biiliocristal) Width in Inches				
Hindu, Delhi	1,157	1.2	973	0.3
United States black	224	1.3	150	0.4
United States white	630	1.4	475	0.4
Polish, Lublin	252	1.4	248	0.5

2. Among United States white infants, from birth to age 6 months twins grow more rapidly in weight and length than singleton infants. Expressed in relation to mean size at birth, average increments for singleton and twin infants are 136% and 183% in body weight and, in body length, 33% and 36% respectively. On the twins, weight and length means at birth are 5.6 pounds and 18.8 inches (2.54 kg and 47.8 cm).

3. During the first postnatal year, Hindu infants gain more slowly than United States black infants, or white infants in Europe and the United States. Compared with United States black infants, average gains for Hindu infants are less by about 3.5 pounds in body weight, 1.5 inches in body length, and over an inch in both head girth and chest girth.

Body weight, on average, more than doubles during the first six months after birth; increments for groups in Table 10 are 118% on Hindu infants, 120% on French infants, and 132% on United States black infants. For the same groups, corresponding increases in body length are 29%, 32%, and 34%. Passing to the period between ages 6 and 12 months, the slowing of infancy growth rates is indicated by average gains on Hindu, French, and United States black infants of 27%, 30% and 32% for body weight, and 11%, 12%, and 13% for body length.

Relative gains in average head girth for the first six months after birth are 21% on Hindu infants, 26% on Polish infants, and 31% on United States white twins. For the same period, gains in chest girth are 25% on Hindu infants, and 35% on United States black and white infants. Comparable values for hip width are 36% and 45% on Hindu and Polish infants respectively. In the second half of the first postnatal year, average gains for United States black and white infants are near 7% in head girth, 8% in chest girth, and 9% (black) to 10% (white) in hip width.

Measures of arm girth taken on United States white infants at Denver, Iowa City, and Newark yield average increases of 1.6 inches, or 42%, between birth and age 6 months, and 0.4 inch, or 8%, between ages 6 and 12 months. Corresponding increments for calf girth of white infants measured at Denver, Iowa City, Newark, and Philadelphia are 2.4 inches, or 56%, in the first 6 months, and 0.5 inch, or 8%, in the second 6 months.

Physically normal, healthy, well-nourished infants of a given ethnic group do not all grow at similar rates. Individuals of northwest European ancestry, fed an adequate diet and given pediatric health supervision from birth, vary in increase of body length between birth and age 6 months from less than 5.5 inches to more than 7.5 inches. Two infants—offspring of mothers given pregnancy care, alike in sex and body length at birth, belonging to the same racial group, and each receiving sound parental and professional guidance—may differ in body length at age 1 year by an amount exceeding 1.5 inches. On the whole, in groups where the gene pool is moderately stable and nurture reasonably satisfactory, individuals who are short at birth may be short, average, or considerably above average at age 1 year; conversely, individuals who are long (tall) at birth may be long, average, or fairly short at age 1 year. The correlation coefficient, r, for body lengths at birth and age 1 year is .5, yielding a forecasting efficiency index near 14%.

Average body length and weight of several human populations at age 1 year. Listed in Table 11 are means for body length and body weight on groups of infants age 1 year studied since 1960 in many parts of the world. Particulars complementing several rows of the table are as follows:

Zambian, Kitwe—the fathers of these infants were copper mine employees. Means for body length and weight on 110 African black infants measured at rural and urban locations in Nigeria are 27.8 inches and 18.1 pounds.

Hindu, Delhi—means for 500 Hindu infants living in rural regions around Poona are lower than the Delhi means by 0.5 inch and 0.5 pounds.

Costa Rican, rural areas—means on 415 Costa Rican urban infants studied at six provincial capitals are higher by 1.0 inch and 0.9 pounds.

Romanian, rural areas—means from measures on 700 Romanian infants residing at urban centers are higher by 0.5 inch and 1.1 pounds.

Russian, Minsk—these means are lower than those on 207 Russian infants at Murmansk by 0.7 inch in length and 2.0 pounds in weight.

Surinam Creole, rural—for body length, the Creole rural mean is the same as that on 635 Creole urban infants. The urban infants are heavier than their rural age peers by 1.3 pounds.

United States black—these means represent infants of middle class families living at Oakland, California. Means on 533 United

States black infants living in widely scattered "poverty areas" are 29.2 inches and 20.5 pounds.

United States white—this sample is comparable to that in Table 11 for United States black infants: both groups are infants of middle class families living under similar economic circumstances. Means are 29.4 inches and 21.1 pounds for 213 white "poverty area" infants.

Table 11

Averages for Vertex-Soles Length in Inches and Body Weight in Pounds at Age 1 Year on Groups of Infants of Both Sexes Measured Since 1960

Group	Number Measured	Mean Body Length	Mean Body Weight
Pakistani, East Pakistan	125	26.0	15.3
Javanese, rural Java	226	26.6	16.6*
Ladino, rural Guatemala	341	26.9	17.0
Zambian, Kitwe	163	27.6	19.0
Hindu, Delhi	973	27.7	17.0
Eskimo, rural Alaska	287	28.0	23.6
Sardinian, Sassari region	992	28.3	20.0
Azerbaijani, rural district	338	28.4	20.3
Costa Rican, rural areas	218	28.5	19.7
Australian aborigine	196	28.6	18.6*
Spanish, rural areas	1,476	28.6	20.3
Romanian, rural areas	880	29.0	20.5
Russian, Minsk	268	29.0	21.6
Chinese, Hong Kong	297	29.1	18.7
Surinam Creole, rural areas	202	29.3	19.5*
Bulgarian, Sofia	612	29.3	21.7*
Turkish, Istanbul	457	29.5	21.6
Italian, Bologna	390	29.5	21.9
United States black	500	29.5	22.3
United States white	1,000	29.6	22.0
Polish, Warsaw	350	29.7	22.5
Tasmanian, Tasmania	1,114	29.9	23.2
Dutch, Netherlands	468	30.0	22.6

* For Javanese, Australian aborigine, Surinam Creole, and Bulgarian groups, the numbers of infants weighed are 868, 330, 379, and 994 respectively.

At age 1 year, from Table 11 and its supporting materials, it is found:

1. Infants of the Netherlands are larger than their contemporary age peers of East Pakistan by 4 inches, or 15%, in average body length, and a little more than 7 pounds, or 45%, in average body weight.

2. Averages for body length are near 27 inches on Ladino infants of mixed Spanish and Amerind ancestry in rural Guatemala, 28 inches on Eskimo infants in rural Alaska, and 29 inches on infants of southern Chinese descent at Hong Kong. Expressed in the metric system, means for body length at age 1 year are near 66 cm on Pakistani infants; 69 cm on Hindu rural infants; 72 cm on Azerbaijani rural infants, and Sardinian infants; and 75 cm on Italian infants at Bologna, Turkish infants at Istanbul, and United States black infants at Oakland.

3. For body weight, averages are near 17 pounds on Hindu infants at Delhi, 20 pounds on Sardinian infants in the province of Sassari, and 23 pounds on Tasmanian infants measured at rural and urban locations. In metric units, body weight means are near 7.5 kg for Javanese infants; 8.5 kg for Australian aborigine infants, and infants of Chinese lineage at Hong Kong; 9.5 kg for Romanian infants at numerous scattered villages; and 10.5 kg for Tasmanian infants.

Averages at age 1 year on 200 Hindu infants reared at Poona under "near optimal" conditions in respect to diet and health care are 28.4 inches for body length and 18.3 pounds for body weight: measures on 1,700 Turkish progeny of well-to-do parents at Istanbul yield means higher by 1.5 inches in length and 4 pounds in weight. Compared with the averages cited earlier on 213 United States white infants living in poverty areas, averages at age 1 year are lower by 0.7 inch and 1.7 pounds on 170 Thai infants at Bangkok showing no sign of malnutrition or deficiency disease.

Average head girth and chest girth at age 1 year. Table 12 presents means at age 1 year for the same variables dealt with at birth in Table 4. Additional averages at age 1 year are cited below:

Hindu, rural—the statistics in this row of Table 12 characterize offspring of parents residing in a rural community near Hyderabad. Mean head girth from measures on 174 middle class Bengali infants at Calcutta is 16.2 inches.

Table 12

Averages for Head and Chest Girths in Inches at Age 1 Year on Infants of Both Sexes Measured Largely Since 1960

Group	Number Measured	Mean Head Girth	Mean Chest Girth
Hindu, rural area	87	16.8	16.0
Hindu, Delhi	973	17.2	17.0
Ladino, rural Guatemala	342	17.2	17.0
Singaporean	210	17.3	------
Tunisian, Chott el Djerid	138	17.4	16.9
United States black	69	17.5	17.2
Thai, Bangkok	130	17.6	17.6
Dominican, Santo Domingo	80	17.6	17.5
Sardinian, Sassari region	992	17.7	18.1
Cambodian, Pnompenh area	200	17.8	17.8
Italian, Pisa province	455	18.1	18.6
Bulgarian, Sofia	565	18.1	18.7
Russian, Kuibyshev	225	18.1	18.7
Japanese, Tokyo	84	18.2	18.2
United States white	1,000	18.3	------
Polish, Lublin	248	18.3	18.8
Guatemalan, well-nourished	282	18.3	19.8
Azerbaijani, rural district	338	18.4	18.8
Polish, Warsaw	230	18.5	19.0
Russian, Kharkov	235	18.6	19.0
Italian, Grosseto province	1,286	18.7	19.3

Singaporean—about 70% of the infants in this sample had Chinese progenitors; the remainder were mainly of Malay and Hindu descent. Means are 17.9 inches for head girth, and 17.6 inches for chest girth, from data on Chinese infants measured in 1955 at Peking.

United States black—the records analyzed in this row were obtained on progeny of low-income families living at Washington, D. C. Averages are higher from other studies. For head girth, means are 18.1 inches on 180 infants of indigent families at New Orleans, and 18.5 inches on 500 offspring of middle class families at Oakland. On 173 Philadelphia infants whose mothers were advised on infant diet and supplied vitamin supplements, means are 18.4 and 18.5 inches for girths of head and chest respectively.

United States white—this mean typifies head girth at age 1 year on infants of Oakland middle-class families. Means are 18.2 and 18.8 inches for head and chest girths of 250 Philadelphia white infants reared under conditions similar to those for the black group described in the preceding paragraph.

It is found:

1. Averages for head girth at the end of the first postnatal year vary from 16.2 inches (41.0 cm) on Hindu infants at Calcutta to 18.7 inches (47.6 cm) on Italian infants in Grosseto province. Intermediate averages are near 17.5 inches on Thai and Tunisian infants; 18.0 inches on Bulgarian infants at Sofia, and Russian infants at Kuibyshev; and 18.5 inches on Polish and Russian infants at Warsaw and Kharkov respectively. The Kharkov average is 2.4 inches (6.1 cm), or 15% higher than that obtained at Calcutta.

2. Differences at age 1 year in average chest girth of infant populations in some instances exceed 3 inches. Compared with the average of 19.8 inches on well-nourished Guatemalan infants, that on Hindu rural infants is smaller by 3.8 inches, or 19%. Chest girth, on average, is near 17 inches for Ladino infants, 18 inches for Sardinian infants, and 19 inches for Polish infants. Part of the difference between chest girth means from any two studies may be due to method, e.g., the measures taken on United States black infants at Philadelphia may have differed slightly from those taken at Washington, D. C. in level of placement or tautness of the tape.

3. For Cambodian, Japanese, and Thai infants age 1 year, average chest girth does not exceed average head girth. On Bulgarian, Italian, and Polish age peers, averages are larger for chest girth than head girth by about 0.5 inch. For a small group of Melanesian infants on the Schouten Islands—14 infants hospitalized with protein-calorie malnutrition, respiratory or digestive disease, anaemia, or symptoms of vitamin deficiency—mean head girth exceeds mean chest girth by 1.0 inch, or 6%.

During infancy healthy infants typically show increase in chest girth relative to head girth. Tables 4 and 12 have several groups in common. For Ladino infants chest girth is 92% of head girth at birth, and 99% of head girth at age 1 year. Corresponding relative increases are from 95% to 102% for Sardinian infants, 96% to 103% for Polish infants, and 97% to 102% for Azerbaijani infants.

**Average stem length, lower limb length, and skelic index at age
1 year.** Means for stem (vertex-rump) length are near 17.0 inches
on Hindu infants at Delhi, and Tunisian village infants living
around Lake Chott el Djerid; 17.5 inches on United States white
infants at Philadelphia; and 18.5 inches on Japanese infants at
Tokyo, offspring of Papiamento-speaking parents residing on Aruba
Island, and Jamaican infants in families of Afro-black ancestry and
low to middle socioeconomic status. Lower limb length is near 10.5
inches on Hindu, Japanese, Jamaican, and Tunisian groups, and
12.0 inches on United States black and white groups.

Lower limb length grows faster than stem length during the first
postnatal year. For Jamaican infants, the skelic index (lower limb
length in percentage of stem length) rises from 51 at birth to 58
at age 1 year. Comparable rises are from 54 to 61 on Hindu infants
at Delhi, and from 48 to 69 on United States white infants at
Philadephia.

Individual and sex differences in body size at age 1 year. There
are small sex differences at age 1 year in most measures of external
body size. Averages for male infants are higher than those for
female infants by about 1.0 pound in body weight, 0.6 inch in body
length and chest girth, 0.5 inch in head girth, and 0.2 inch in hip
width, arm girth, and calf girth.

In body weight, individuals age 1 year vary from less than 10 to
32 pounds among Creole infants of rural Surinam, and from 16 to
34 pounds (7.3 to 15.4 kg) among healthy, well-nourished United
States white infants.

Among Ladino infants of rural Guatemala body weight is under
14 pounds for 1 child in 10. For 10% in each instance, body weight
is below 15 pounds on Nigerian village infants receiving pediatric
health supervision; 16 pounds on Costa Rican urban infants; 17
pounds on United States black and white infants living in poverty
areas; 18 pounds on United States white twins; and 19 pounds
on Dutch and Swedish infants. The heaviest 10% of infants in these
groups have body weights at or above 20 pounds (Ladino, Nigerian), 23 pounds (Jamaican), 24 pounds (Costa Rican, United
States twins), and 25 pounds (Dutch, Swedish).

Variation in body length at age 1 year extends from 22 to 36
inches among Angolan black infants of the Lunda and Tshokwe
tribes, 23 to 38 inches among infants of Hindu descent in rural
Surinam, and 25 to 36 inches (63.5 to 91.5 cm) among United States

white infants living in urban communities. One infant in 10 of the groups named is shorter than 25.5 inches (Ladino), 26.5 inches (Costa Rican, Nigerian), 27.5 inches (United States twins, and poverty area residents), or 28.5 inches (Dutch, Swedish). Another 1 in 10 is longer than 28.5, 29.5, and 31.0 inches among Ladino, Nigerian, and Dutch infants respectively.

In other body dimensions, measures on United States healthy infants of northwest European ancestry show individual infants dispersed from 17 to 23 inches (43 to 58 cm) for stem length, 17 to 20 inches for head girth, 16 to 21 inches for chest girth, 5.5 to 9.0 inches for calf girth, 4.5 to 7.5 inches for arm girth, and 4.0 to 7.0 inches (10 to 18 cm) for hip width.

On the United States white population age 1 year, standard deviation estimates are 2.5 pounds for body weight; 1.2 inches for body length; 0.9 inch for chest girth; 0.7 inch for stem length; 0.5 inch for girths of head, arm, and calf; and 0.3 inch for hip width.

Suggested Readings

Hansman, C. Anthropometry and related data, in R. W. McCammon (Ed.), **Human Growth and Development.** Springfield, Ill.: C. C. Thomas, 1970 (101-154).

Karlberg, P., I. Engstrom, H. Lichtenstein and I. Svennberg **Acta Paediatrica Scandinavica,** 1968, 187, 48-66 (Supplement).

Kasius, R. V., A. Randall, W. T. Tompkins and D. G. Wiehl **Milbank Memorial Fund Quarterly,** 1958, 34, 355-362.

Kornfeld, W. **Oesterreichische Zeitschrift für Kinderheilkunde und Kinderfürsorge,** 1954, 10, 71-88.

*Meredith, H. V. **Child Development,** 1970, 41, 551-600.

*Meredith, H. V., and A. W. Brown **Human Biology,** 1939, 11, 24-77.

Wilson, R. S. **Annals of Human Biology,** 1974, 1, 175-188.

CHAPTER VI

Ages of Children at Oral Emergence of Primary Teeth

Entrance of teeth into the mouth is part of the process of dental eruption: for each tooth, the entire eruption process involves movement out of a crypt of bone, through soft tissues, and continuing until the crest of the tooth reaches the occlusal (biting) level. Age of emergence is defined as the stage when any piece of tooth enamel has perforated the gum and can be seen on inspection of the oral cavity.

Primary teeth are referred to frequently as deciduous teeth, and occasionally as lactal or milk teeth. The primary dentition consists of 20 teeth: proceeding from the midline of the face, each side of each jaw holds a central incisor, lateral incisor, canine (cuspid), primary first molar, and primary second molar.

Age at which oral emergence of the first tooth occurs. Average age at emergence of the first tooth is 7.2 months for 180 British infants at Newcastle-on-Tyne; 7.3 months for 171 Swedish infants at Umea; 7.5 months for 2,060 United States white infants at Iowa City, New York, Philadelphia, and Washington, D. C.; 7.9 months for 300 Czech infants at Prague, and 8.1 months for Hungarian infants at Debrecen and Hajdusamson. On other groups, averages are 7.1 months for 2,010 United States black infants at New York and Philadelphia; 7.3 months for 520 United States mixed black and white infants at Detroit and New York; 7.7 months for 220 Hutu infants in Rwanda, and 50 Melanesian infants at villages on the northwest coast of New Guinea; and 8.0 months for 89 Tutsi infants in Rwanda. It follows that in most human populations studied to date, average time of beginning tooth emergence is between the postnatal ages of 7 and 8 months.

Means for initial tooth emergence in selected white subgroups are 8.1 months for 80 pairs of twins, 9.1 months for 29 sets of triplets, 12.2 months for a set of quintuplets, 7.4 months for 55 offspring of mothers having rubella during pregnancy, and 12.3 months for 134 infants with Down's syndrome (Kalmuc idiocy, Mongolism).

Rarely does a tooth appear in the oral cavity during the first third of the year following birth. There are instances of tooth emergence prior to birth, and in the first postnatal month. Emergence of a tooth between ages 1 and 4 months is documented for an Australian aborigine infant at 1 month; a United States white infant at 1.5 months; a Canadian white infant at 2 months; a Pima Amerind infant at 2.5 months; a German twin, Korean, Melanesian, and United States black infant at 3 months, and an African Mandinka infant at 3.5 months.

Sometimes tooth emergence does not begin prior to the second postnatal year. Examples are absence of any tooth emergence for an Australian aborigine infant age 14 months, an African Mandinka infant age 15 months, a United States white singleton infant age 16 months, a mentally deficient infant age 18 months, a triplet age 19 months, a white infant age 21 months whose mother had rubella during pregnancy, and a child age 3 years with Down's syndrome.

A study on 390 Punjabi rural children varying in age from birth to 3 years finds no child to have an emerged tooth before age 8 months: a study on 1,290 United States black infants at ages from birth to 1 year finds 60% show tooth emergence before 8 months. An emerged tooth at ages between 6 and 9 months is found for 15% of 102 Bengalese infants, and 65% of 1,410 New York white infants.

For white nonpathologic infants from singleton and twin births, emergence of a first tooth occurs usually during the last two-thirds of the first postnatal year, and almost always before age 16 months. The first tooth to emerge is commonly a lower central incisor, but occasionally is an upper central incisor, upper lateral incisor, or primary first molar. There is no sex difference in the average age primary tooth emergence is initiated.

Age of infants at oral emergence of different primary teeth. The usual sequence of emergence for human primary teeth is incisors, first molars, canines, and second molars. Within the incisor group, the typical order is lower central, upper central, upper lateral, and lower lateral. Averages for age at emergence of different primary teeth are displayed in Tables 13 and 14.

Several rows of Tables 13 and 14 are based on seriatim (longitudinal) records. Sample size is 116 for Swedish infants in Norrbotten county; 84 for British infants at London; 208 for Canadian twins at Toronto; 171 for Swedish infants at Umea; 300 for United States white infants, some residing at Denver and others at or near

Yellow Springs; and 43 for white infants with Down's syndrome at Melbourne.

Table 13

Mean Age in Months at Oral Emergence of Primary Incisor Teeth in Children of Both Sexes

Group	Central Incisors		Lateral Incisors	
	Lower	Upper	Upper	Lower
Swedish, Norrbotten County	7.0	8.8	9.8	11.4
British, London	7.3	9.2	10.6	11.5
Hutu, Rwanda	7.8	9.4	10.6	11.7
Canadian twins, Toronto	7.6	9.5	10.5	12.1
Tutsi, Rwanda	8.2	9.8	11.4	12.1
Melanesian, Bougainville Island	8.8	9.5	11.5	12.3
Korean, Korean province	7.7	9.8	12.3	12.4
Swedish, Umea	7.6*	9.8*	10.9*	12.7*
United States white	7.5*	9.3*	11.4*	13.3*
Melanesian, Siassi Islands**	8.2	9.4	11.3	13.4
Bundi, New Guinea	8.3	10.8	13.1	13.4
Australian, Melbourne		10.1*	12.0*	13.6*
Hungarian, three cities	8.7*	10.5*	12.3*	14.3*
French, Seine Department	8.1	10.8	12.1	14.6
Canadian quintuplets	13.3	12.9	15.6	15.8
Down's syndrome, Melbourne	13.7*	18.5*	23.3*	23.3*

* These averages are computed from median emergence ages for the right and left antimeres.
** Three hundred individuals were examined at ages from 5 to 24 months.

In other rows, means are based on records for children examined on one occasion. Sample size varies from 140 to 222 for Hutu children, 49 to 90 for Tutsi children, 73 to 101 for Melanesian village children on Bougainville Island, and 440 to 1,400 for Korean children in the province of Chulla-Book-Do. The total number of children examined is near 500 for Australian normal children at Melbourne, 3,000 for French residents of "département de la Seine," and 1,100 for Hungarians at Debrecen, Miskole, and Mátészalka.

Table 13 shows:

1. For British, French, Hungarian, Swedish, Australian white, and United States white infants, average age of tooth emergence is

between 7.0 and 8.7 months for the lower central incisor, 8.8 and 10.8 months for the upper central incisor, 9.8 and 12.3 months for the upper lateral incisor, and 11.4 and 14.6 months for the lower lateral incisor.

2. The average age at which upper and lower lateral incisors emerge is 1 to 2 months earlier for Hutu infants in Rwanda than for French infants in the department of Seine, and Bundi infants in the New Guinea highlands.

3. Compared with average age of incisor emergence in non-pathologic infants, average age of emergence in infants with Down's syndrome is later by at least 5 months for the central incisors and 8 months for the lateral incisors.

From Table 14, it is found:

1. The primary first molars, on average, emerge between 15 and 16 months after birth. Excluding the Korean group, average ages are between 18 and 20 months for the primary canines. Averages for the primary second molars are more variable, falling between 25 and 29 months on Australian, European, and North American white groups.

2. There are differences of 4 months for the primary first molar and 7 months for the primary second molar teeth, between average emergence ages in Bundi children of New Guinea and Australian white children with Down's syndrome. Average age at emergence of the primary canines is about 3 months earlier for Korean than French infants.

In a sample of 1,900 white children—living in New York families of fairly low economic status, and under regular nutritional guidance and medical supervision—the youngest ages at which primary teeth emerged are 4 months for the lower central incisor, 5 months for the upper central incisor, 6 months for the upper and lower lateral incisors, and 8 months for each of the canine, first molar, and second molar teeth. By contrast, the earliest emergence times for the Dionne quintuplets are 11 months for an incisor tooth, 17 months for a first molar tooth, 22 months for a canine tooth, and 34 months for a second molar tooth. In the New York sample, the oldest ages without oral emergence of given primary teeth are 15 months for the upper central incisor, 17 months for the lower central incisor, 21 months for the upper lateral incisor, 27 months for the lower lateral incisor and the first molars, 29 months for the canines, and 34 months for the second molars.

Table 14

Mean Age in Months at Oral Emergence of Primary Canine and Molar Teeth in Children of Both Sexes

Group	First Molars*	Canines*	Second Molars*
Bundi, New Guinea	15.5	18.0	22.7
Korean, Korean province	15.8	16.2	23.9
Melanesian, Siassi Islands	16.5	19.0	24.5
Hungarian, three cities	16.0	19.7	25.0
Swedish, Norrbotten County	15.3	18.4	25.5
British, London	14.8	18.3	26.0
Melanesian, Bougainville Island	16.1	18.9	27.3
Swedish, Umea	15.6	19.0	27.6
French, Seine Department	16.1	19.4	------
United States white	16.1	20.0	28.3
Australian, Melbourne	16.0	20.2	28.4
Down's syndrome, Melbourne	19.5	34.3	30.4
Canadian quintuplets	19.0	24.2	36.8

* Average of means for corresponding teeth in the upper and lower jaws.

Number of primary teeth visible in the mouth at selected infancy ages. Exhibited in Tables 15 and 16 are averages for number of primary teeth emerged at ages 6 months, 1 year, 18 months, and 2 years. Comment relating to several rows of these tables follows:

Mandinka, Gambia—the statistics are for infants living at four villages about 100 miles (160 km) from Bathurst. From records obtained at another Gambian village means are 4.5, 10.9, and 17.4 at ages 1, 1.5, and 2 years respectively.

Hindu, rural—means from records on infants residing at villages around Hyderabad.

United States white—these means are derived from data amassed at Boston, Iowa City, New Haven, New York, Yellow Springs, and Washington, D. C. On Detroit infants, about 70% white and 30% black, means are 0.4 and 5.9 for number of emerged teeth at ages 6 months and 1 year.

Tunisian, rural—the infants are progeny of parents living at villages around Chott El Djerid, a lake in southern Tunisia.

United States black—these means represent healthy infants in families of low to middle socioeconomic status at Washington, D. C.

Table 15

Average Number of Primary Teeth Showing Exposure of Enamel (Oral Emergence) in Infants Age 6 months and 1 Year

Group	Age 6 Months		Age 1 Year	
	Number of Infants	Mean Number of Teeth	Number of Infants	Mean Number of Teeth
Mandinka, Gambia	105	0.2	95	3.9
Australian Aborigine	34	0.1	29	4.0
Hindu, rural	65	0.1	33	4.2
Chimbu, New Guinea	------	----	76	5.2
Shambaa, Tanzania	------	----	50	5.3
Swiss, Zurich	81	0.4	81	5.3
Uganda black, Kampala	101	0.3	71	5.4
Senegal black, Dakar	246	0.5	263	5.5
French, Paris	294	0.4	257	5.8
United States black	------	----	530	5.9
Chinese, Singapore	------	----	103	5.9
British, London	174	0.4	159	6.1
United States white	740	0.5	698	6.2
Czech, Prague	------	----	186	6.1
Scottish, Aberdeen	------	----	126	6.2
Swedish, Umea	------	----	171	6.4
Tunisian, rural	------	----	140	7.1

Tables 15 and 16, in conjunction with related findings, show:

1. The average infant age 6 months lacks oral emergence of any tooth. This is found in human populations of Africa, Asia, Australia, Europe, and North America. For white infants, about 1 in 3 has between 1 and 6 emerged teeth.

2. At age 9 months, the typical number of emerged teeth is 3, and variation among individuals is from 0 to 9. For United States white infants of each sex separately, the average number of emerged teeth is 3.4 for males, and 2.7 for females.

3. Means for number of primary teeth emerged at age 1 year are

near 4 on Australian aborigine, African Mandinka, and Asian Hindu infants; near 5 on Chimbu, Shambaa, and Swiss infants in New Guinea, Africa, and Europe; and near 6 on British, Chinese, Czech, French, North American black and white, and Scottish infants. On American white infants of each sex, means indicate oral emergence of primary teeth is about 2 weeks earlier for males than females. Individual differences extending from 0 to 20 emerged teeth are seen among white, black, and Chinese infants. About 98% of United States white infants age 1 year have between 2 and 12 emerged teeth.

Table 16

Average Number of Primary Teeth Showing Exposure of Enamel (Oral Emergence) in Infants Age 18 Months and 2 Years

| Group | Age 18 Months | | Age 2 Years | |
	Number of Infants	Mean Number of Teeth	Number of Infants	Mean Number of Teeth
Shambaa, Tanzania	25	10.8	23	15.2
Australian Aborigine	42	12.1	34	15.9
Swiss, Zurich	81	12.2	81	16.2
Hindu, rural	40	10.0	44	16.3
British, London	147	12.9	116	16.3
Finnish, Finland	------	------	193	16.4
French, Paris	172	12.3	123	16.4
Swedish, Umea	171	12.8	171	16.4
United States white	538	12.4	480	16.7
Czech, Prague	295	12.9	323	16.8
Mandinka, Gambia	84	10.0	89	17.0
Scottish, Aberdeen	------	------	126	17.4
Senegal black, Dakar	230	12.3	164	17.5
Chimbu, New Guinea	50	13.1	34	17.7
Uganda black, Kampala	37	13.2	18	18.0
Tunisian, rural	114	13.5	100	18.0

4. At age 18 months, in most human populations, the average number of primary teeth in the oral cavity falls between 12 and 13. Averages are about 10 teeth for some groups of African black and Hindu infants, and near 13 for other African black groups and

some European groups. Rarely does a physically normal infant
have less than 4 or more than 18 erupted teeth at age 18 months,
although instances of 0 and 20 are seen. Among 230 black infants
at Dakar, all had between 4 and 18 teeth; and among 240 white
infants at Boston, 97% had between 6 and 16 teeth.

5. The average number of primary teeth emerging at or before
age 2 years commonly is between 16 and 18. Groups for which this
is true include Australian aborigines; African blacks in Gambia,
Senegal, and Uganda; Europeans at Aberdeen, London, Paris,
Prague, Umea, and Zurich; New Guinea Chimbu; and United States
whites at Boston, Iowa City, New York, and Yellow Springs. Among
individuals age 2 years, oral emergence usually has occurred for
no less than 6 teeth, and may have occurred for the entire primary
dentition. Studies on United States white infants show about 15%
with 8 to 14 teeth, 60% with 15 to 18 teeth, and 25% with 19 to
20 teeth. In a British study at London on 116 infants age 2 years,
11% were found to have 20 emerged teeth.

**Age and speed at which emergence of the primary dentition is
completed.** Mean age at oral emergence of the last primary tooth
in 170 Swedish children at Umea is 2.5 years. From a study on
United States black and white children, means are earlier by 2
months.

For about 90% of United States white children, the last primary
tooth emerges between ages 2 and 3 years. The earliest times re-
corded for emergence of the terminal primary unit are 10 months
in a Canadian white child; 1 year in a Chinese child, and United
States black child; 17 months in an Australian aborigine and a
white quintuplet; 18 months in a Gambian black child; 19 months
in an Apache Amerind child; and 31 months in a child with Down's
syndrome. In a group of 210 white Boston children age 3 years,
4% had less than 20 emerged teeth. There are records of a quintuplet and a child with Down's syndrome lacking oral emergence
of the completed series of primary teeth at age 40 months.

For Swedish children (and probably other white children) the
average time between oral emergence of the first and last primary
teeth is 22 months. Oral emergence occurs more slowly in some
individuals than others: in about 90% of children duration of the
emergence period is between 17 and 30 months. There are instances
of 20 teeth emerging in 9 months for an American black child, 12

months for an American white child, 15 months for a British child, and 37 months for an American white child.

Suggested Readings

Boutourline, E., and G. Tesi **Human Biology,** 1972, 44, 433-442.

*Leighton, B. C. **Practitioner,** 1968, 200, 836-842.

Lysell, L., B. Magnusson and B. Thilander **Odontologisk Revy,** 1962, 13, 217-234.

McGregor, I. A., A. M. Thomson and W. Z. Billewicz **British Journal of Nutrition,** 1968, 22, 307-314.

*Meredith, H. V. **Journal of Tropical Pediatrics and Environmental Child Health,** 1973, 19, 195-199.

*Roche, A. F., and D. H. Barkla **Journal of Mental Deficiency Research,** 1964, 8, 54-64.

Tasanen, A. **Annales Paediatriae Fenniae,** 1969, 14, Supplement 29, 1-40.

CHAPTER VII

Standing Height and Body Weight in Early Childhood

Early studies on average height and weight of children at ages from 3 to 5 years were published in 1836 in Belgium, between 1860 and 1879 in England and Italy, and between 1880 and 1885 in Germany and the United States.

Height and weight of several human populations at age 4 years. Presented in Table 17 are averages for standing height and body weight of children age 4 years living in Africa; Asia; Australia; Europe; North, Central, and South America; the islands of Aruba and New Guinea; and the Phillipine Islands. Each average is for children of both sexes measured since 1960.

Notations on some of the groups in Table 17 will precede drawing comparisons among the groups:

Tibetan, Nepal—the sample comprises 42 children measured at a lowland settlement, and 31 at highland settlements.

Hindu, rural—members of low income families residing in rural regions around Hyderabad. Subgroup averages on 1,050 children lacking any symptom of nutritional deficiency are 34.7 inches (88 cm) in height, and 25.8 pounds (11.7 kg) in weight.

Bolivian, southwest Bolivia—this group included children of Amerind (largely Aymara and Quechua), Spanish, and Cholo (mixed Amerind-Spanish) ancestry.

Hawaiian, Oahu—children participating in feeding programs at day care centers; about 40% "Hawaiian-Caucasian," 40% "Orientals and Filipinos," and 20% white, black, or Samoan.

Russian, urban centers—the urban centers are Baku, Blagoveshchensk-on-Amur, Cherboksary, Kalinin, Kazan, Kursk, Orenburg, Petrozavodsk, Pskov, Ryazan, and Saratov.

United States white—data obtained between 1963 and 1967 at Baltimore and Birmingham, and during 1968-1973 in Washtenaw County, Michigan, 34 other states, and the District of Columbia.

United States black—children measured during 1963-1967 at Baltimore, Birmingham, New Orleans, and Washington, D. C.; 1968-

[53]

1973 in Washtenaw County, Michigan, 26 other states, and the District of Columbia.

Table 17

Averages for Body Weight in Pounds and Standing Height in Inches on Children of Both Sexes Age 4 Years Measured Since 1960

Group	Number Measured	Mean Body Weight	Mean Body Height*
Tibetan, Nepal	73	22.3	33.1
Hindu, rural area	1,525	25.6	34.5
Bundi, New Guinea	95	28.7	34.8
Lunda, rural Angola	1,799	------	35.5
Bolivian, southwest Bolivia	134	30.9	35.8
Filipino, Luzon and Visayas	453	27.3	36.0
Shi, rural Zaire	333	29.0	36.0
Ladino, rural Guatemala	308	30.3	36.0
Spanish, rural villages	1,628	33.7	38.5
Costa Rican, rural and urban	675	32.8	38.7
Chinese, Taiwan	1,993	33.4	38.9
Bulgarian, rural and urban	831	35.5	39.3
Surinam Creole	1,526	34.2	39.4
Hawaiian, Oahu	255	35.7	39.4
Italian, six provinces	1,175	35.7	39.7
Russian, urban centers	3,374	35.8	39.7
United States white	1,184	35.5	39.9
United States black	1,646	35.7	40.3
Aruba islander	109	36.3	40.3
Australian, Sydney	2,974	37.7	40.3
Belgian, Liège province	504	36.4	40.4
Latvian, Riga	200	37.7	40.5
Czech, Prague	398	37.7	40.8
Dutch, Netherlands	1,401	38.6	41.3

* On average, vertex-soles distance is about 0.6 inch (1.5 cm) less in the standing position than in the position of dorsal recumbency.

From Table 17, it is found that at age 4 years:

1. Averages are less for Tibetan children than for contemporary Dutch children by about 8 inches (20.5 cm) in standing height, and 16 pounds (7.4 kg) in body weight.

2. Czech children at Prague surpass Hindu rural children living in the vicinity of Hyderabad by about 6 inches in average standing height, and 12 pounds in average body weight. From 49 children of East Pakistan, among whom malnutrition was prevalent, means are 34.8 inches for height, and 24.5 pounds for weight.

3. Means for standing height are near 36 inches (91 cm) on Filipino children, African black children of the Shi tribe, and Ladino (mixed Spanish and Amerind) children in Guatemala. Australian white children at Sydney are taller by 4 inches, or 11%.

4. Compared with Angolan black children of the Lunda district, Spanish rural children are taller by 3 inches, Surinam Creole children taller by almost 4 inches, and Latvian children at Riga taller by 5 inches. The Latvian urban children are 4 pounds, or 12%, heavier than the Spanish rural children.

There is wide variation in average standing height at age 4 years among Amerind tribes. Means are 34.5 inches on 400 Maya village children in Guatemala, 36.0 inches on 25 Quechua children in the Peruvian highlands, 37.4 inches on 97 semi-nomadic Amerind children in Surinam, and 39.5 inches on Blackfeet, Assiniboine, and Gros Ventres children in Montana.

Height and weight of socioeconomic and racial groups within cities. Assembled in Table 18 are averages for standing height and body weight on young children of both sexes whose families differ greatly in socioeconomic status. At the cities of Baghdad, Bogotá, Bombay, Ibadan, and Santiago, children in the 'low' category are from homes designated poorest, underprivileged, or poverty level; and those in the 'high' category from homes designated well-to-do, high income, or culturally advantaged. The Istanbul parents are "unskilled factory workers" or "wealthy," and the Delhi parents unskilled, or in professional and major managerial positions. The other groups juxtaposed are English children attending schools in slum districts, or elite private schools; and United States children attending Philadelphia schools in economically "underprivileged" or "favored" districts.

Table 18 shows:

1. Young children in upper class families, on average, are taller and heavier than those in lower class families. This is seen consistently for the African, Asian, European, North and South American, and Middle East cities represented.

Table 18

Standing Height in Inches and Body Weight in Pounds for Young Children of Parents in High and Low Socioeconomic Categories

Group	Children in Lower Class Homes		Children in Upper Class Homes	
	Number	Mean	Number	Mean
Children Age 4 Years Measured Since 1960: Height				
Iraqi, Baghdad	38	38.2	89	39.3
Hindu, Bombay and Delhi	240	35.4	50	39.5
Turkish, Istanbul	117	36.0	70	39.6
Colombian, Bogotá	69	36.9	64	39.7
Chilean, Santiago	120	37.2	35	40.8
Yoruba, Ibadan	234	37.8	219	40.8
Children Age 6 Years Measured During 1945-52: Height				
United States white*	72	44.9	72	46.0
British, several towns	107	43.4	48	46.2
Children Age 4 Years Measured Since 1960: Weight				
Colombian, Bogotá	69	30.1	64	34.3
Turkish, Istanbul	119	30.6	70	34.7
Hindu, Bombay and Delhi	240	26.5	50	35.7
Yoruba, Ibadan	278	30.8	234	35.9
Iraqi, Baghdad	38	32.6	89	36.6
Chilean, Santiago	120	32.2	35	37.3
Children Age 6 Years Measured During 1945-52: Weight				
British, several towns	107	42.8	48	46.7
United States white*	72	45.2	72	48.5

* Philadelphia.

2. At Istanbul and Santiago, averages for standing height of young children living in well-to-do homes exceed those of their age peers residing in poor homes by about 3.5 inches, or 9 cm. Socioeconomic differences near 3 inches are shown on Colombian children at Bogotá, Nigerian children at Ibadan, and British pupils at "elite" and "slum area" schools. The difference of 1.1 inches for

United States children at Philadelphia is similar to that of 0.9 inch obtained at Eugene, Oregon, from measures taken in 1950 on boys age 7 years, 51 having fathers in "unskilled and semiskilled" occupations, and 55 in "professional and major managerial" occupations.

3. Compared with average body weight of young children in low income families, young children in high income families are heavier by between 4 and 5 pounds at Baghdad, Bogotá, Ibadan, Istanbul, and Santiago. The smaller difference of 3.3 pounds on children age 6 years at Philadelphia is consistent with that of 2.4 pounds on boys age 7 years at Eugene.

At age 5 years, several averages are available from measures of standing height taken concurrently on two or more racial groups living in the same city. Data collected between 1968 and 1970 at Johannesburg yield means higher for 133 white children than 208 black children by 2.5 inches. From measures taken in the late 1930's at Los Angeles, means are 42.0 inches for 111 American Japanese children, 42.6 inches for 703 American Mexican children, 43.7 inches for 135 American black children, and 44.0 inches for 2,810 American white children. In contrast with the Johannesburg height differences between black and white groups, the Los Angeles black and white groups are similar in average height.

Records obtained during 1961-1964 at four cities in the Soviet Union show average standing height at age 5 years is similar for Kirghiz and Russian children at Frunze, while at Baku, Cherboksary, and Kazan, Russian children are taller than Azerbaijani, Chuvash, and Tatar children by 0.4 to 0.9 inch. Sample size is between 210 and 400 for each group.

Urban-rural similarities and differences in height and weight of young children. Table 19 brings together averages for height and weight of children age 4 years measured largely since 1960 in rural areas and at urban centers. Successive rural groups in the upper section of the table are from villages in Bombay province, rural areas of Surinam, collective farms in the Kanashsky region of the Soviet Union, the village of Imesi in western Nigeria, rural areas of Surinam, widely scattered Bulgarian villages, isolated Polish villages in the Ostroleka and Suwalki districts, and a Czechoslovakian rural region near Olomous. The corresponding urban centers are Delhi, Surinam cities, Cheboksary, Lagos, Surinam cities, Sofia, Warsaw and Prague.

Table 19

Averages at Age 4 Years for Height and Weight of Rural and Urban Children of Both Sexes Measured Largely Since 1960

Group	Rural Children		Urban Children	
	Number	Mean	Number	Mean
Standing Height in Inches				
Hindu	1,525	34.5	310	36.8
Surinam Indonesian	99	38.1	47	38.1
Chuvash	219	37.3	198	38.4
Nigerian	250	37.8	60	38.8
Surinam Creole	404	39.3	622	39.4
Bulgarian	182	39.1	304	39.9
Polish	77	38.5	280	40.1
Czech	124	40.1	398	40.8
Body Weight in Pounds				
Hindu	1,525	25.6	310	28.7
Surinam Indonesian	100	31.2	48	31.5
Nigerian	250	31.3	60	33.6
Chuvash	219	32.6	198	33.8
Surinam Creole	408	34.3	626	34.2
Bulgarian	182	35.1	304	36.5
Polish	77	34.8	280	36.7
Czech	124	36.5	398	37.7

Averages for standing height show:

1. In the Chuvash, Hindu, and Polish comparisons, urban children exceed rural children by 1.1 to 2.3 inches.

2. In the Bulgarian, Czech, and Nigerian comparisons, urban children exceed rural children by 0.7 to 1.0 inch.

3. In the Surinam Creole and Indonesian comparisons, urban and rural children are practically alike. Means for standing height at age 4 years from a large English study made during 1909-1910 are 37.7 inches on 8,930 rural children, and 0.6 inch lower on 6,880 urban children.

A companion set of statements for average body weight is as follows:

1. Urban Hindu, Nigerian, and Polish young children are heavier than their respective rural age peers by about 2 to 3 pounds.

2. Bulgarian, Chuvash, and Czech urban children weigh between 1.0 and 1.5 pounds more than their ethnic peers living in rural districts.

3. Surinam Creole and Indonesian rural and urban children do not differ significantly in body weight. From the English study cited above, average body weight is 34.1 pounds on rural children, and 0.6 pound less on urban children.

In summary, urban-rural differences in mean body weight of young children are (1) neither consistently in one direction, nor found in all groups, and (2) more often higher for children living in urban communities than those inhabiting rural regions.

Height and weight of young children in relation to birth weight and twin birth. On average, children weighing 5.5 pounds or less at birth are shorter and lighter in early childhood than those weighing over 5.5 pounds at birth. At age 4 years, means for height and weight of children with birth weights at or below 5.5 pounds are 36.0 inches and 27.0 pounds on 83 Hindu children living at Delhi, and—on 72 United States black children living in poverty areas— 39.5 inches and 33.6 pounds. Children with birth weights exceeding 5.5 pounds are larger by 1.1 inches and 2.6 pounds for 246 children at Delhi, and by 0.8 inch and 2.5 pounds for 334 United States black "poverty area" children.

Means at age 4 years from measures on 368 United States white twins residing in the metropolitan Louisville area are 39.5 inches for standing height and 34.5 pounds for body weight. On a contemporary group of 200 children this age described as representative of the United States white population, means are higher by 0.7 inch and 1.6 pounds for height and weight respectively.

Height and weight of young children living at different altitudes. From studies at different elevations in the Soviet Union and Japan it is found, on average, young children reared at high altitudes are smaller than those reared nearer sea level. At age 5 years, means for standing height and body weight are 40.4 inches and 34.8 pounds on 197 Kirghiz children at Naryn (elevation 6,600 feet, or 2.0 km), and higher by 1.7 inches and 4.3 pounds on 272 Kirghiz children at Dzhalal-Abad (elevation 2,300 feet, or 0.7 km). On Japanese children of Kagawa province, means at age 5 years are

40.4 inches and 34.8 pounds for 360 inhabitants of the mountainous region, and higher by 0.6 inches and 1.6 pounds for 690 inhabitants of the coastal region.

Height and weight of young children in relation to health status and health care. Findings are available from several studies in the 1960's on children of similar health status. Means for standing height and body weight at age 4 years are 37.5 inches and 29.0 pounds on 70 "healthy" Malayan children, 38.5 inches and 32.1 pounds on 142 Jamaican children showing "no unequivocal signs of malnutrition," 39.1 inches and 32.1 pounds on 126 Hindu children of Agra province with no history of "illness likely to retard growth," 39.3 inches and 32.6 pounds on 283 "well nourished" Thai children of middle and upper class families, 40.8 inches and 35.9 pounds on 219 Nigerian children at Ibadan raised in "elite" homes. These groups differ in average height by 3.3 inches (8.4 cm), and in average weight by 6.9 pounds (3.1 kg): the averages for Malayan, Jamaican, and Thai children do not exceed the averages of 39.4 inches and 35.5 pounds for contemporary United States black children living in poverty areas.

Studies conducted in Guatemala, Nigeria, and the Soviet Union report averages for height and weight of young children receiving adequate to superior health guidance and care. At a Maya village in the Guatemalan highlands, the mothers of 180 children were given periodic advice on the feeding of their children, and provided with supplemental foods; at a Yoruba village in Nigeria, the parents of 250 children were furnished "the best medical services possible to any sick child," and supplied food supplements for any child showing signs of malnutrition; and at Petropavlovsk, 175 Russian children were given "superior" postnatal care. Height and weight averages at age 4 years are 35.6 inches and 29.5 pounds for the Maya children, 37.8 inches and 31.3 pounds for the Yoruba children, 40.3 inches and 37.4 pounds for the Russian children. Comparatively: (1) the Maya averages exceed by 1.3 inches and 1.5 pounds those on 230 Maya children living at a highland village where no program of health or nutritional guidance was sponsored, (2) the Russian averages exceed by 0.6 inch and 2.3 pounds whose on 164 Petropavlovsk children reared under inadequate conditions of care, (3) the Yoruba averages are lower than those for children of Ibadan upper class families by 3.0 inches and 4.6 pounds, and (4) averages for the Maya group furnished nutritional guidance are lower than those for contemporary Amerind children of Montana by 3.9 inches (10 cm) and 8.0 pounds (3.6 kg).

The assortment of findings in the preceding two paragraphs indicates the need for extensive further research directed toward clarifying, in different settings, relative contributions of genetic heritage, disease prevention, dietary intake, health hygiene, and exercise regimen to body size in early childhood.

Sex and individual differences in height and weight of young children. Averages for standing height at successive ages in early childhood are identical for girls and boys in some populations and, in other populations, higher for boys than girls by amounts up to 0.5 inch. Limiting attention to investigations at age 5 years on large samples of children (over 200 at each age), averages are practically alike from studies on Angolan Lunda, East German, Filipino, Surinam Creole, and Hungarian boys and girls, and higher for boys than girls by 0.3 to 0.5 inch in studies on Belgian, Bulgarian, Chinese, Dutch, Finnish, Hindu, Japanese, Spanish, and United States black and white children.

From the same group of studies, averages for body weight at age 5 years are similar (differ less than 0.5 pounds) on girls and boys of East Germany, Finland, and Hungary, and are higher on boys than girls by 0.5 to 1.5 pounds for Belgian, Bulgarian, Chinese, Dutch, Filipino, Hindu, Japanese, Surinam Creole, and United States black and white children. In the aggregate, at age 5 years human males exceed females by 0.3 inch (0.7 cm) in standing height, and slightly less than 1.0 pound (0.4 kg) in body weight.

For height and weight, there are individual differences in rate of growth during early childhood. This is a commonplace for groups varying widely in genetic background, socioeconomic status, and health care, but less commonly recognized in regard to groups fairly homogenous racially and culturally. Therefore, differences will be cited among United States children of northwest European ancestry, born into homes of middle to upper socioeconomic status, well-nourished from birth, and under continuing pediatric health care. Annual gains in height within this group—between ages 4 and 5 years—are less than 2.3 inches for 10% of individuals, 2.3 to 2.5 for 20%, 2.6 to 2.8 inches for 40%, 2.9 to 3.1 inches for 20%, and more than 3.1 inches for 10%. Between ages 5 and 6 years, variation in height increments is about the same.

Continuing with the group specified, annual increases in body weight from ages 4 to 5 years are below 3.2 pounds for 10% of individuals, 3.2 to 4.2 pounds for 20%, 4.3 to 5.8 pounds for 40%, 5.9 to 7.3 pounds for 20%, and above 7.3 pounds for 10%. Increases in body weight between ages 5 and 6 years are slightly higher than between 4 and 5 years.

In body size at age 4 years, physically nonpathologic children differ as much as 20 inches in standing height and 50 pounds in body weight. Among 890 Lunda children in northwest Angola, the shortest individual is 28 inches (71 cm): among 2,900 Australian children at Sydney, the tallest individual is 48 inches (122 cm). The lowest body weight among 680 Surinam children of Hindu and Pakistani ancestry does not exceed 15 pounds (6.8 kg): the heaviest individual among 1,000 Surinam Creole children weights 65 pounds (29.5 kg).

Table 20 illustrates individual differences at age 4 years by giving, for several groups, the height and weight values below which 10% of each group fall, and above which another 10% fall. The table shows:

1. Fewer than 10% of Dutch children are under 39 inches in height, or 33 pounds in weight. In contrast, 10% of Ladino children are more than 5 inches shorter, and 7 pounds lighter.

2. Over 10% of Australian children at Sydney are above 42.5 inches in height, and 44.5 pounds in weight. Among Yoruba rural children in Nigeria, 9 in 10 are below 40 inches in height, and 36 pounds in weight.

3. Standing height at age 4 years is less than 38.5 inches for 10% of Belgian children at Courtrai, and more than 90% of Ladino children in rural Guatemala.

4. At Hong Kong, among children with Chinese progenitors, slightly more than 10% weigh over 35 pounds; at Sydney, among children with northwest European progenitors, this weight is exceeded by 74% of the group.

5. In most human populations, 80% of children age 4 years have standing heights within bounds of 4 to 5 inches. Bounds at this age that include 80% of body weights are about 9 pounds apart for rural children in Jamaica and Nigeria, and 12 pounds apart for Australian, Italian, and United States black children.

Individual differences in the standing height of young children are validly described by the standard deviation. At age 5 years, it

Table 20

Percentiles for Standing Height in Inches and Body Weight in Pounds at Age 4 Years on Children of Both Sexes Measured Since 1960

Group	Number Measured	Tenth Percentile	Ninetieth Percentile
Standing Height			
Ladino, rural Guatemala	308	33.9	38.2
Yoruba, rural Nigeria	250	36.0	40.0
Singaporean	275	36.3	40.6
Jamaican, rural	320	36.5	40.6
United States white*	217	36.7	41.9
United States mixed**	287	38.0	42.1
United States black*	530	37.5	42.5
Italian, Emilia department	292	37.6	42.7
Belgian, Courtrai	942	38.4	42.7
Australian, Sydney	2,980	38.2	42.8
Dutch, Netherlands	1,400	39.2	43.4
Body Weight			
Ladino, rural Guatemala	307	26.5	34.4
Chinese, Hong Kong	201	26.6	35.2
Singaporean	277	26.3	35.9
Yoruba, rural Nigeria	250	26.9	35.9
Jamaican, rural	585	27.6	36.5
Antilles black, St. Kitts	818	27.2	37.0
United States white*	218	29.7	40.6
Belgian, Courtrai	942	31.4	41.9
United States black*	470	30.6	42.6
Italian, Emilia Department	292	30.9	43.0
Dutch, Netherlands	1,400	33.5	43.9
Australian, Sydney	2,970	32.5	44.8

* Children living in poverty areas.
** United States Health and Nutrition Examination Survey conducted 1971-1974.

has been shown: (1) distributions of standing height closely approximate the Gaussian model, (2) within a given population, standard deviations are practically alike for females, males, and both sexes together, and (3) standard deviations are near 1.8 inches (4.6 cm) for Chinese, Czech, Dutch, Hungarian, and Japanese children, near 2.0 inches for Australian, Belgian, Chuvash, and German children, and near 2.4 inches (6.1 cm) for African Lunda, Asian Hindu, and South American Creole children. Distributions for body weight are slightly skewed to the right in early childhood, and become increasingly skewed (elongated by heavy body weights) during the childhood years.

Body size of young children in relation to size at birth and in late childhood. Knowledge of a child's overall body size at birth is of little value for predicting the child's body size in early childhood. Pearsonian r's are near .3 for (1) birth weight with weight at age 5 years, and (2) birth length with height at age 5 years. These weak positive associations differ substantially from the moderately strong positive associations of overall body size in early and late childhood: r's are near .9 for (1) height at age 5 years with height at age 9 years, and (2) weight at age 5 years with weight at age 9 years. Lacking information on a child's size at a given age, the "best guess" of size at a later age would be the average for the later age. Improvement on this—with the aid of measurements, r's, and regression equations—is less than 5% in forecasting height or weight at age 5 years from length or weight at birth, and more than 50% in forecasting height or weight at age 9 years from comparable measures at age 5 years. Examples of these relations in non-statistical language are (1) a child who is short at birth may be short, average, or tall at age 5 years, (2) a child who is heavy at birth may be heavy, average, or light at age 5 years, and (3) a child who is tall at age 5 years will likely be tall or moderately tall at age 9 years.

Suggested Readings

Jones, D. L., and W. Hemphill **Height, weight and other physical characteristics of New South Wales children: Part II. Children under five years of age.** Sydney: Health Commission of New South Wales, 1974.

Low, W. D. **Zeitschrift für Morphologie und Anthropologie,** 1971, 63, 11-45.

Meredith, H. V. Selected anatomic variables analysed for interage relationships of the size-size, size-gain, and gain-gain varieties, in

L. P. Lipsitt and C. C. Spiker (Eds.), **Advances in child development and behavior.** New York: Academic Press, 1965. (Volume 2, 221-256).

*Meredith, H. V. Research between 1960 and 1970 on the standing height of young children in different parts of the world, in H. W. Reese and L. P. Lipsitt (Eds.), **Advances in child development and behavior.** New York: Academic Press, 1978. (Volume 12, 1-59).

Meredith, H. V., and E. M. Meredith **Child Development,** 1950, 21, 141-147.

*Tanner, J. M., R. H. Whitehouse and M. Takaishi **Archives of Disease in Childhood,** 1966, 41, 613-635.

Yarbrough, C., J-P Habicht, R. M. Malina, A. Lechtig and R. E. Klein **American Journal of Physical Anthropology,** 1975, 42, 439-447.

CHAPTER VIII

Size and Form of Head, Trunk, and Limbs in Early Childhood

Head width, face width, and cephalic index. Syntheses of studies on United States white children age 4 years indicate average head width is 5.4 inches (13.8 cm), and face width 4.3 inches (10.9 cm). These averages are based on 1,500 measures of head width (bi-parietal diameter) and 660 measures of face width (bizygomatic diameter). Individual United States white children vary from 4.7 to 6.1 inches in head width, and 3.8 to 4.8 inches in face width. About 10% of children have head widths less than 5.0 inches and face widths less than 4.0 inches, while another 10% have head and face widths exceeding 5.8 and 4.6 inches respectively.

Average head width at age 4 years is near 5.0 inches for children on the islands of Aruba and Sardinia; 5.4 inches for Belgian, British, Finnish, and German children; 5.6 inches for Apache Amerind, Czech, and Hungarian children; and 5.8 inches for Buryat children at Siberian villages in the Soviet Union. Averages for face width are near 4.4 inches on Belgian, Finnish, and German children; 4.6 inches on Apache and Czech children; and 4.8 inches on Buryat children. Females have slightly (no more than 0.2 inch) narrower head and face widths than males.

Compared with average head and face widths for United States white infants at birth, children age 4 years are larger by 1.8 inches (4.5 cm) in width of head, and 1.4 inches (3.6 cm) in width of face.

Between ages 3 and 6 years, size of the head and face increases slowly. Average increases for the entire triennium are 0.2 inch in head width and 0.3 inch in face width.

The head is wide in relation to width of face, and narrow in relation to distance from front (glabella) to back of the head. On average, at age 4 years head width varies from 120% of face width on 65 Buryat children, through 124% on 455 Belgian children, to 127% on 660 United States white children. Comparably, width of the head in percentage of its anteroposterior dimension (the cephalic index) averages 75 for Aruba children; near 81 for Belgian, Finnish, Swiss, and United States white children; and 90 for Apache and Buryat children.

[67]

Head girth, chest girth, and cephalo-thoracic index. Listed in Table 21 are averages for head and chest girths at age 4 years. Particulars on several of the samples are as follows:

Hindu, rural—children lacking any symptoms of nutritional deficiency residing in low income rural families around Hyderabad. Averages are slightly lower, 18.4 inches for each girth, on 467 Hindu village children showing signs of vitamin deficiency or protein-calorie malnutrition.

Hindu, Delhi—children from homes of all socioeconomic levels.

United States black—members of low income families living in the District of Columbia.

German, East Germany—physically normal children representative of the entire Deutschen Demokratischen Republik.

Azerbaijani, Chuvash, Kirghiz—children of three Turkic-speaking minority groups in republics of the Soviet Union.

Italian, two provinces—members of families living in the east-central provinces of Grosseto and Pisa.

From Table 21 it is found:

1. Russian urban children age 4 years exceed Hindu rural children by about 1.5 inches, or 8%, in average head girth, and almost 2.0 inches, or 10%, in chest girth.

2. Averages for head girth are between 19.5 and 20.0 inches on Australian white, Bulgarian, Czech, Finnish, French, Italian, and Swiss children.

3. For chest girth, averages are between 20.5 and 21.0 inches on Bulgarian, Finnish, German, Polish, and Sardinian children; they are between 21.0 and 21.5 inches on Azerbaijani, Chuvash, Czech, French, Hungarian, Italian, Kirghiz, and Tatar children.

Measures at age 4 years taken during 1956-57 on Swiss children at Basel, and during 1950-53 on United States white children at Newark, New Jersey, yield similar averages. Those on 100 children at Newark are 19.6 inches for head girth, and 20.2 inches for chest girth: compare these with corresponding values in Table 20 on children at Basel.

Within white populations—and probably within all human populations—young children vary less, and grow more slowly, in head girth than chest girth. Among Australian white children at Sydney, head girths at age 4 years are smaller than 19.0 inches (48.2 cm) for 10% of children, and larger than 20.5 inches (52.0 cm) for 10%. Growth in head girth between ages 3 and 6 years averages slightly less than 0.4 inch each year.

Table 21

Averages for Head and Chest Girths in Inches on Groups of Children of Both Sexes Age 4 Years Measured Between 1955 and 1975

Group	Number Measured	Mean Head Girth	Mean Chest Girth
Hindu, rural	1,058	18.5	18.6
Hindu, Delhi	310	19.1	19.4
United States black	96	19.3	19.7
Aruba islander	109	19.5	------
Australian, Sydney	1,798	19.7	------
Swiss, Basel	129	19.7	20.4
Bulgarian, rural and urban	1,668	19.8	20.8
German, East Germany	1,949	------	20.8
Finnish, rural and urban	209	19.9	20.8
Polish, Warsaw	200	------	20.8
Sardinian, rural and urban	150	19.5	20.9
Azerbaijani, Baku	281	------	21.0
Chuvash, rural and urban	417	------	21.1
Czech, Prague	400	19.9	21.1
Hungarian, Budapest	473	------	21.2
Tatar, Kazan	214	------	21.2
French, Paris	216	19.9	21.3
Kirghiz, Frunze	210	------	21.3
Italian, two provinces	564	19.9	21.5
Russian, Lvov and Saratov	487	20.1	21.5

Individual differences in size and growth rate of chest girth will be detailed for United States white children reared in homes of middle and upper socioeconomic status by parents taking advantage of resources for pediatric health care and advice on child nutrition. At age 4 years, chest girths are smaller than 19.7 inches (50.0 cm) for 10% of these children, 19.7 to 20.2 inches for 20%, 20.3 to 21.2 inches for 40%, 21.3 to 21.9 inches for 20%, and larger than 21.9 inches (55.6 cm) for 10%. Comparable values at age 6 years are 20.6 inches or less for the smallest 10%, 21.3 to 22.4 inches for the central 40%, and 23.4 inches or more for the largest 10%. Increases in chest girth between ages 4 and 5 years are no greater than 0.2 inches for the 10% of children having the slowest growth, 0.4 to 0.8 inches for the average gainers, and 1.2 inches or more for the 10% increasing most rapidly. Averages over the period from 3 to 6 years are slightly higher each year, rising from about 0.5 inch between ages 3 and 4 years, to about 0.8 inch between ages 5 and 6 years.

Chest girth, on average, is larger than head girth in early childhood. The two series of averages in Table 21 for Bulgarian, Czech, Finnish, French, Italian, Russian, and Swiss children show that typically the chest girth of white children age 4 years is larger than head girth by 1.2 inches (3 cm).

Chest girth in percentage of head girth (the cephalo-thoracic index) increases during infancy and early childhood. For Hindu children at Delhi, chest girth is 95% of head girth at birth (Table 4), 99% at age 1 year (Table 12), and 102% at age 4 years (Table 21); corresponding values for Sardinian children are 95%, 102%, and 107%. Average indices on Italian children of Grosseto province are 95 at birth, 103 at age 1 year, 105 at age 3 years, and 111 at age 6 years; practically identical indices characterize Bulgarian children at Sofia.

Sitting height, lower limb height, and skelic index. Averages for sitting height at age 4 years vary from near 20 inches to slightly more than 23 inches. Means at or near 20.0 inches (50.8 cm) are shown in Table 22 on children of the Bundi tribe measured in 1967 at villages in the Madang district of New Guinea, and Hindu children measured between 1965 and 1967 at rural locations in the vicinity of Hyderabad. On 285 United States white children of northwest European ancestry measured in the 1930's at Boston, and during 1946-55 at Iowa City, the mean is 23.2 inches (59.0 cm).

Several groups studied in Europe, including the Belgian, Bulgarian, Finnish, Hungarian, and Swiss groups depicted in Table 22, have average sitting heights between 22.5 and 23.0 inches.

United States black children, on average, have a shorter body stem (sitting height) than white children. Table 22 shows the mean on United States black children in economically poor families is 0.6 to 1.2 inches lower than the means on European white children. Means for sitting height on United States children age 6 years measured between 1974 and 1977 in Richland County, South Carolina, are lower for 420 black children than 120 white children by 0.3 inch (0.8 cm).

Table 22

Averages in Inches for Sitting Height and Lower Limb Height of Children of Both Sexes Age 4 Years Measured Between 1955 and 1975

Group	Number Measured	Mean Sitting Height	Mean Lower Limb Height
Hindu, rural	1,058	19.9	14.8
Bundi, New Guinea	96	20.0	14.8
Hindu, Delhi	310	21.4	15.4
Bulgarian, rural and urban	1,642	22.5	16.8
French, Paris	218	22.3	16.9
Finnish, rural and urban	209	22.5	17.1
Belgian, Uccle	73	22.8	17.2
Hungarian, Budapest	473	22.9	17.4
Aruba islander	109	22.8	17.5
Swiss, Basel	129	22.7	17.8
United States black*	96	21.7	18.3

* Children of low income families living in the District of Columbia.

Among the 473 Hungarian children entered in Table 22, the shortest and longest body stems are 19.7 and 26.4 inches. Notice that the shortest Hungarian child and the average Bundi child are similar in this dimension. The sitting heights of well-nourished United States white children age 4 years are spread with 10% below 22.2 inches, and 10% above 24.2 inches.

Between ages 4 and 5 years, increases in sitting height of well-nourished white children are 0.8 inch or less for 10%, 0.9 inch for 20%, 1.0 to 1.2 inches for 40%, 1.3 inches for 20%, and 1.4 inches or more for 10%. Throughout early childhood there is a slow decline in growth rate: average gain in sitting height is near 1.3 inches between ages 3 and 4 years, and 1.0 inch between ages 5 and 6 years.

As shown in Table 22, averages for lower limb height (standing height minus sitting height) at age 4 years are 14.8 inches (37.6 cm) for Hindu rural children; near 17.0 inches for Belgian, Bulgarian, Finnish, and French children; and 17.5 inches (44.5 cm) for children living on Aruba Island. United States black children have longer lower limbs than Bundi children by about 3.5 inches (1.4 cm), or 24%.

Among well-nourished United States white children age 4 years, individuals differ in lower limb height to the extent that 10% have limbs shorter than 16.5 inches, and 10% limbs longer than 18.6 inches. At age 6 years, differences in lower limb height within and between the two South Carolina groups referred to previously are revealed by the following: for white children, there are 10% below 18.7 inches, 20% between 18.7 and 19.4 inches, 40% between 19.5 and 20.5 inches, 20% between 20.6 and 21.3 inches, and 10% above 21.3 inches; for black children, there are 10% below 19.7 inches, 40% between 20.6 and 21.6 inches, and 10% above 22.5 inches. The shortest and longest lower limbs among the 420 United States black children are 17.4 inches (44.3 cm) and 24.7 inches (62.8 cm) respectively.

During the age period between 4 and 6 years, annual increments in lower limb height of United States white children are less than 1.3 inches (3.3 cm) for 10%, 1.3 to 1.4 inches for 20%, 1.5 to 1.7 inches for 40%, 1.8 to 1.9 inches for 20%, and more than 1.9 inches (4.8 cm) for 10%.

On average, at age 4 years lower limb height is 74% of sitting height for Bundi village children in New Guinea, and 84% of sitting height for United States black children at Washington, D. C. Skelic indices between 76 and 77 typify Finnish, French, Hungarian, and Aruba children age 4 years. The relatively long lower limbs of black children are evident from the South Carolina study at age 6 years: on black children, the lowest 10% of skelic indices fall between 73 and 80, and the highest 10% between 90 and 95; comparable indices on white children are systematically less, spreading from 68 to 75 and 83 to 87 for the lower and upper 10%.

Shoulder width, hip width, and shoulder-hip indices. Shoulder width (biacromial diameter) of white children age 4 years varies for different individuals from 6 inches to 12 inches, and averages about 9 inches. Means from specific studies are 8.7 inches on 455 Belgian children at Brussels; 8.9 inches on 473 Hungarian children at Budapest, and 1,670 Bulgarian rural and urban children; 9.1

inches on 350 United States white children at Boston, Denver, and Iowa City; and 9.2 inches on 2,880 Australian children at Sydney. Among Australian and United States white children, shoulder widths are narrower than 8.6 inches for 10%, and wider than 9.9 inches for 10%.

Increments in shoulder width between ages 4 and 5 years are less than 0.3 inch for 1 child in 10, between 0.3 and 0.8 inch for 8 in 10, and more than 0.8 inch for 1 in 10. Between ages 5 and 6 years, increments are similar. For white children of Australia, Europe, and North America, average increase in shoulder width between age 3 years and age 6 years is 1.4 inches.

Means at age 4 years from studies of hip width (biiliocristal diameter) are 6.0 inches on 150 Sardinian children; 6.1 inches on 310 Hindu children at Delhi; 6.2 inches on United States black children at Washington, D. C.; 6.5 inches on 455 Belgian children at Brussels; 6.7 inches on 209 Finnish children, 473 Hungarian children at Budapest, and 489 United States white children at Boston, Denver, and Iowa City; and 6.8 inches on 1,670 Bulgarian children.

The narrowest 10% of hip widths among United States white children age 4 years are between 5.7 and 6.2 inches, and the widest 10% between 7.2 and 7.7 inches. At age 6 years, the narrowest 10% of hip widths are between 5.6 and 6.3 inches for United States black children, and between 6.3 and 6.9 for United States white children. The widest 10% fall between 7.4 and 8.1 inches in the white group, and between 7.9 and 8.5 inches in the black group.

Among United States white children, gains in hip width between ages 4 years and 5 years are under 0.3 inch for 10%, between 0.3 and 0.5 inch for 80%, and above 0.5 inch for 10%. Average increase in hip width between age 3 years and age 6 years is about 1.0 inch.

The hips relative to the shoulders are narrower for young children of black ancestry than for those of white ancestry. At age 4 years, average hip width is 69% of average shoulder width on 96 United States black children at Washington, D. C. The same hip-shoulder index is 74 on 455 Belgian children at Brussels, and 330 United States white children at Iowa City; 75 on 473 children at Budapest, and 150 United States white children at Denver; and 77 on 1,670 Bulgarian children. Stated in terms of reciprocals, at age 4 years shoulder width relative to hip width is 145% for the black sample, and between 131% and 135% for the white samples. Average indices of girls and boys are similar.

Arm girth, calf girth, upper limb length, and limb indices. The arm girth dealt with here is measured midway between shoulder and elbow, with the upper limb relaxed and the measuring tape lightly contacting the skin.

At age 4 years, averages for this measurement are 5.3 inches (13.4 cm) on 467 Hindu rural children showing symptoms of malnutrition; 5.4 inches on 1,058 Hindu rural children lacking any sign of nutritional deficiency; 5.8 inches on 94 black children living at villages in southern Malawi, and on 50 Sudanese children residing in a rural area south of Khartoum; 6.3 inches on 1,665 Bulgarian children, 455 Belgian children at Brussels, and 129 Swiss children at Basel; 6.4 inches on 242 United States white children at Denver, Colorado, and Newark, New Jersey; 6.5 inches on 218 French children at Paris, and 280 United States white children at Iowa City; and 7.0 inches (17.7 cm) on 255 Hawaiian well-nourished children of Oriental, Filipino, and white ancestry living on Oahu Island in low to middle income families.

Among United States white children receiving adequate nutritional and health care, arm girth at age 4 years is less than 6.1 inches (15.5 cm) for 10%, between 6.4 and 6.9 inches for 40%, and greater than 7.4 inches (18.8 cm) for 10%. At age 6 years, 10% of the arm girths on 2,000 British children measured in 1959 at London are smaller than 6.0 inches, and 10% larger than 7.5 inches. Among 420 United States black children age 6 years measured during 1974-77 at Columbia, South Carolina, arm girths are less than 6.3 inches for 10%, and greater than 7.8 inches for 10%.

For white children of the United States, increase in average arm girth between ages 3 and 6 years is about 0.6 inch. During any given year, 1 child in 10 shows no gain, and 1 in 10 gains 0.4 inch or more.

Averages for calf girth (maximum circumference of the leg in the calf region) at age 4 years are 6.8 inches (17.3 cm) on 467 Hindu rural children giving indications of malnutrition; 7.0 inches on 1,058 Hindu rural children manifesting no symptom of inadequate nutrition; 8.2 inches on 129 Swiss children at Basel; 8.6 inches on 280 United States white children at Iowa City, and 218 French children at Paris; and 8.7 inches (22.1 cm) on 149 United States white at Portland, Oregon.

Among the 280 well-cared-for Iowa City children, calf girths at age 4 years are between 7.5 and 7.9 inches for 10%, and between 9.4 and 9.9 inches for 10%. At age 6 years, calf girths on 420 United States black children fall between 7.3 and 8.4 inches on 10%, and between 10.1 and 11.8 inches on 10%; small and large girths at this age on the 2,000 London children are below 8.6 inches or above 10.3 inches respectively.

For healthy white children of the United States, increases in calf girth between ages 4 and 5 years are 0.1 inch or less for 10%, 0.2 inch for 20%, 0.3 or 0.4 inch for 40%, 0.5 inch for 20%, and 0.6 inch or more for 10%. Increase in average calf girth from age 3 years to age 6 years is about 1.1 inches.

Upper limb length is measured from the shoulder (acromiale) to the tip of the middle finger, with the limb extended at the side of the body. Measures taken since 1960 in Belgium and Bulgaria yield identical means at age 4 years (16.8 inches, or 42.7 cm) on 488 Belgian and 1,668 Bulgarian children. From data collected during 1937-46 on 115 United States white children having parents largely in the professional and managerial classes, average upper limb length is 17.1 inches. Within this group, the 10% with the shortest upper limbs fall between 15.1 and 16.2 inches, and the 10% with the longest upper limbs between 18.1 and 19.4 inches.

In the United States study, average upper limb length at age 6 years exceeds that at age 3 years by 3.7 inches (9.4 cm), or almost 24%. Means at age 6 years in the Belgian and Bulgarian studies are higher than those at age 4 years by 2.4 and 2.3 inches respectively.

Continuing with the United States white group: typically, arm girth at age 4 years is slightly less than 40% of upper limb length. Among individuals, indices below 40 register amount of upper limb slenderness, and indices above 40 amount of upper limb stockiness. Indices below 37 identify the slimmest 10% of upper limbs, and those above 43 the stockiest 10%.

Averages at age 4 years for calf girth in percentage of lower limb height are 46 on 129 Swiss children at Basel, 50 on 229 United States white children at Iowa City, and 51 on 218 French children at Paris. Indices on the Iowa City group are below 46 for 10%, and above 54 for 10%. At age 6 years, indices on 420 United States black children at Columbia, South Carolina, vary from 35 to 40 for 10%, average 44, and vary from 48 to 58 for 10%. Indices at this age on 400 Columbia and Iowa City white children show 10% below 42, the average at 46, and 10% above 50. It follows that during early childhood the lower limbs of United States white children become more slender with age, and at any given age are more stocky than the lower limbs of United States black children.

Several studies on United States white children reared under favorable conditions of nutrition and health supervision have investigated the influence of childhood illnesses on growth in stem height, lower limb height, chest girth, shoulder width, hip width, arm girth, and calf girth. These studies show no relation in early childhood between number or kind of non-hospitalized illness episodes and body size, physique, or growth rates during semi-annual and longer periods.

Boys, on average, are slightly larger than girls during early childhood in each of the body dimensions discussed in this chapter.

Suggested Readings

Banik, N. D. D., R. Krishna, S. I. S. Mane, L. Rai and A. D. Taskar Indian Journal of Medical Research, 1970, 58, 135-142.

Eiben, O., G. Hegedüs, M. Bánhegyi, K. Kiss, M. Monda and I. Tasnádi Budapesti Óvodások és iskolások testi fejlettsége. Budapest: Eötvös Loránd University, 1971.

Meredith, H. V. Growth, 1947, 11, 1-50.

Meredith, H. V. Journal of Pediatrics, 1950, 37, 183-189.

Meredith, H. V. Growth, 1954, 18, 111-134.

*Meredith, H. V. Child Development, 1968, 39, 335-377.

*Meredith, H. V., and V. B. Knott American Journal of Diseases of Children, 1962, 103, 146-151.

Twiesselmann, F. Développement biométrique de l'enfant a l'adulte. Bruxelles: Presses Universitaires de Bruxelles, 1969.

CHAPTER IX

Standing Height and Body Weight in Late Childhood

Average height and weight of human populations at age 9 years. Listed in Table 23 are averages from measures of standing height and body weight taken since 1960 on groups of children age 9 years living at hamlets, villages, or other rural and semi-rural locations. The statistics in row 19 on Russian children of the Soviet Union are from data collected in the Altai territory, the Buinsk and Vitebsk areas, rural regions of the Kola peninsula, and rural locations in the vicinity of Kalinin, Kiev, Kirovograd, Moscow, Pskov, Ryazan, Stavropol, and Ulyanovsk.

Table 23 shows that at age 9 years:

1. Averages for standing height of Sherpa and Bundi highland children in Nepal and New Guinea are about 9 inches (23 cm) lower than the average on Blackfeet children in northwest Montana. For body weight, the mean on Blackfeet children exceeds that on Sherpa children by about 26 pounds, or 70%.

2. Thai and Hindu village children, on average, are 5 inches (12.7 cm) shorter and 17 pounds (7.7 kg) lighter than Lithuanian children residing in the Moletsk region of the Lithuanian Soviet Socialist Republic. There is a similar difference in standing height between means on black children living in the Lunda district of Angola, and at coastal villages of Guyana.

3. Means for height near 48.5 inches typify Chuvash, Moldavian, Tatar, and Turkish rural children. Corresponding means for weight are between 53 and 56 pounds.

Average standing height is the same for Fijian village children (Table 23), 165 Sudanese children measured during 1964-1968 at villages south of Khartoum, and 583 Creole children measured during 1964-1965 in rural areas of Surinam. Mean body weight of Fijian children is 6 pounds higher than for Sudanese children, and almost 2 pounds higher than for Surinam Creole children.

Means at age 9 years for standing height and body weight of contemporary groups of city children are presented in Table 24. Notations on this table follow:

[77]

Table 23

Average Standing Height in Inches and Body Weight in Pounds for Children of Both Sexes Age 9 Years Measured Since 1960 in Rural and Semi-rural Areas

Group	Number Measured	Mean Body Weight	Mean Body Height
Sherpa, Nepal highlands	47	37.3	43.3
Bundi, New Guinea highlands	90	43.0	43.3
Thai, scattered villages	54	42.6	45.4
Hindu, Palghar Taluk region	834	43.1	45.4
Lunda, northeast Angola	1,639	------	45.6
Zapotec, southern Mexico	40	47.7	46.6
Quechua, Peru highlands	78	51.4	46.9
Tajik, south Tajik S.S.R.	208	50.3	47.3
Kisi and Nyakyusa, Tanzania	71	50.6	48.0
Chuvash, Kanashsky region	256	53.9	48.3
Moldavian, Moldavian S.S.R.	476	55.9	48.4
Turkish, Etimesgut area	225	53.0	48.5
Tatar, northeast Tatar S.S.R.	194	53.2	48.7
Gypsy (Romi), Czechoslovakia	115	56.0	49.1
Bushnegro, Surinam interior	171	56.4	49.9
Costa Rican, rural regions	182	55.5	50.0
Bulgarian, broad rural sample	363	58.4	50.2
Lithuanian, Moletsk area	229	60.0	50.4
Russian, 12 rural regions	4,613	58.7	50.5
Fijian, coastal villages	152	57.9	50.6
Cree, James Bay area	82	62.3	50.9
Guyanese black, coastal	438	56.7	51.0
Blackfeet, northwest Montana	88	63.5	52.3

Hindu, Coimbatore—these averages are for girls only. Averages obtained at Delhi on children age 9 years "in normal health on clinical examination" are 45.2 pounds and 48.2 inches for 322 girls, and higher by 1.4 pounds and 0.4 inches for 281 boys. Using these differences, estimated means for children of both sexes at Coimbatore are 44.6 pounds and 47.0 inches.

Brazilian, Salvador—subgroup means are 50.5 pounds and 48.0

inches on 143 "white and light mulatto" children, and higher by
2.5 pounds and 1.3 inches on 180 "dark mulatto and black" children.

Russian, 26 cities—children residing at cities within the extensive
zone from Murmansk in the north, to Vladivostok in the east,
Stavropol in the south, and Pskov in the west. Averages vary from
59.8 pounds and 50.0 inches on 252 children at Cheboksar to 67.0
pounds and 52.1 inches on 772 children at Lvov.

Canadian, three cities—children living at Halifax, London, and
Montreal. Means on the 563 children measured at Montreal—all
French-Canadians—are 59.5 pounds and 50.7 inches for height and
weight respectively.

Polish, four cities—the cities are Danzig, Lódz, Lublin, and Warsaw.

Dutch, three cities—these cities are Amsterdam, Nijmegen, and
Utrecht.

From Table 24, and complementary studies, it is found:

1. German children at Düsseldorf are, on average, 6 inches (15
cm) taller, and 20 pounds (9 kg) heavier than Hindu children at
Coimbatore. Means on 5,500 children measured in 1960 at several
cities in Chile are 48.6 inches and 55.8 pounds; compared with the
tabled means on United States black children at Columbia, South
Carolina, these values are lower by 4 inches (10 cm), and 10
pounds (4.5 kg).

2. At Manila, Cap Bon, Norilsk, and Oslo, averages for standing
height are near 47, 49, 51, and 53 inches respectively. At Manila,
Hong Kong, Cap Bon, Gorki, and Tiflis, sequential averages for
body weight are near 45, 50, 55, 60, and 65 pounds.

3. Average height at age 9 years is 49.5 inches for Chinese children at Hong Kong, and Japanese children at Sendai. Sardinian
children at Sassari have about the same height; the mean for 176
Sassari children measured in 1965 is 49.6 inches (126 cm). Means
are almost identical (50.6 or 50.7 inches) for Honduran Creole
children at Belize City, French-Canadian children at Montreal,
445 Costa Rican urban children, and 360 Surinam urban children
of Hindu ancestry. The Costa Rican and Surinam groups were
measured during 1963-1965.

Displayed in Table 25 are averages for height and weight of
children age 9 years representing different countries, provinces,
islands, and other geographical zones having both rural and urban
inhabitants. Explanatory comment follows:

Table 24

Average Standing Height in Inches and Body Weight in Pounds for Children of Both Sexes Age 9 Years Measured Since 1960 at Urban Centers

Group	Number Measured	Mean Body Weight	Mean Body Height
Hindu, Coimbatore*	520	43.9	46.8
Filipino, Manila	1,215	45.3	47.2
Brazilian, Salvador	534	51.7	48.7
Japanese-Brazilian, Bauru	275	53.3	48.8
Chuvash, Cheboksar	240	53.8	49.2
Tunisian, Cap Bon	592	55.1	49.2
Chinese, Hong Kong	670	49.9	49.5
Japanese, Sendai	7,650	55.6	49.5
Moldavian, Kishinev	530	56.9	49.7
Egyptian, Cairo	731	59.7	49.9
Honduran Creole, Belize City	715	57.4	50.6
Russian, 26 cities	11,107	60.7	51.1
Canadian, three cities	1,387	60.7	51.2
French, Paris	210	59.7	51.2
Moçambique black, Lourenço Marques	395	61.5	51.4
Polish, four cities	1,145	62.1	51.4
Bulgarian, Sofia	399	61.7	51.6
Belgian, Brussels and Courtrai	2,129	61.3	51.7
Australian, Perth and Sydney	2,319	63.5	51.7
Georgian, Tiflis	290	65.7	52.1
Hungarian, Budapest	546	64.0	52.3
U. S. black, Columbia, S. C.	485	65.7	52.6
Dutch, three cities	700	64.3**	52.8
Norwegian, Oslo	2,150	65.9	52.9
German, Düsseldorf	3,659	65.6	53.1

* Girls only, see text.
** Amsterdam and Utrecht only, see text.

Philippine region—Luzon Island and the Visayas.

Italian provinces—Aquila, Chieti, Grosseto, Naples, Pisa, Rome, Teramo, and Udine.

Czechoslovakia—means are 59.8 pounds and 51.1 inches for children living in the Slovak districts; 63.5 pounds and 52.0 inches for Czech children in the Bohemian and Moravian districts.

United States—a representative sample of United States children living in the 48 contiguous states, except for children confined to institutions or residing on Amerind reservations.

Taking Tables 23, 24, and 25 together, it is found:

1. In different parts of the world children age 9 years vary in average standing height from less than 44 inches (Table 23) to 53 inches (Table 24). Average body weight varies from below 45 pounds (Table 23) to above 65 pounds (Table 24). Comparatively short height and light weight typify rural groups living in India, Thailand, and New Guinea: relatively tall height and heavy weight characterize urban groups in Hungary, Norway, and West Germany; black, white, and Blackfeet Amerind groups in the United States; and national groups in East Germany, the Netherlands, and New Zealand.

2. Means for standing height near 47 inches are common to Andean Quechua children (Table 23), Hindu children at Coimbatore (Table 24), Filipino and Malayan children (Table 25), and Taiwan aborigine children. The mean for 360 Taiwan aborigines age 9 years measured during 1970-1972 is 47.1 inches. Means near 51 inches are common to Cree children in Canada (Table 23), Russian children at Orel (Table 24), children on the Italian peninsula (Table 25), and Creole urban children in Surinam. The Surinam mean on 857 Creole urban children measured during 1964-1965 is 51.0 inches.

3. Hindu rural children residing in Palghar Taluk, west-central India (Table 23) are shorter and lighter than rural children of Hindu descent living in Surinam by at least 3.5 inches and 5.5 pounds. The Surinam averages are 49.2 inches and 49.0 pounds from measures on 1,970 rural children. Compared with Egyptian children of the United Arab Republic (Table 25), Dutch children of the Netherlands are almost 4 inches taller and about 10 pounds heavier.

Average height and weight at age 9 years of privileged and underprivileged children. Means for standing height and body weight of environmentally advantaged and disadvantaged children are assembled in Table 26. In the upper section of the table:

Table 25

**Mean Standing Height in Inches and Body Weight in Pounds for
Children Age 9 Years Sampled Since 1960 to Represent Countries,
Provinces, Districts, and Islands**

Geographical Zone	Number Measured	Mean Weight	Mean Height
Philippine region	385	46.1	47.1
Malaya	378	46.3	47.1
Southwest Bolivia	307	52.2	47.2
Northeast Brazil	221	50.8	47.3
Zambia, Zambezi district	77	45.8	47.4
Venezuela	304	52.5	48.7
Taiwan, except aborigines	1,293	50.2	49.0
Egypt	21,313	54.2	49.2
Japan	18,349	57.4	50.3
Poland, Brzeziny County	1,393	58.4	50.3
Bulgaria	1,163	61.3	50.5
Barbados Island	972	55.9	50.6
Italian provinces	4,730	60.9	50.8
Uruguay	156	60.3	51.0
Finland	2,043	59.3	51.4
Aruba Island	258	60.0	51.4
Belgium, Liège province	863	59.8	51.6
Czechoslovakia	14,000	61.8	51.6
England, except London	1,159	60.5	51.7
New Zealand, except Maoris	1,816	64.7	52.1
East Germany	2,202	62.8	52.2
United States	1,207	65.0	52.2
Netherlands	1,061	64.4	53.1

1. The Chinese, Dutch, Filipino, and Turkish groups consist of children with at least one parent in a professional or major managerial occupation. Parents of the Filipino children, for instance, are described as "professionals, proprietors, or administrators."

2. Children in the other groups have "wealthy," "prosperous," "well-to-do" parents. Two groups are girls: (1) Hindu girls in Hyderabad families of high socioeconomic status, and (2) Baganda girls attending a residential private school near Kampala. Statistics

on these groups are included since sex differences in average height and weight at age 9 years are slight, and not consistently in one direction. From data collected during 1956-1965 on 7,400 Hindu children, means for males exceed those for females by 0.3 inch in height and 0.4 pound in weight. From data collected during 1974-1977 on 485 United States black children, means for females exceed those for males by 0.3 inch and 1.0 pound.

In the lower section of the table:

1. The Hindu, Nigerian, Peruvian mestizo, and United States groups are from "urban slums," "poverty areas," and "low income" homes; and the Colombian mestizo children are residents of "a subsistence farming community" where "there is chronic protein-calorie malnutrition."

2. For the remaining groups, parents are in either the unskilled occupational category (Chinese, Turkish), or the broader unskilled and semiskilled category (Dutch, Filipino).

Examination of the averages for age 9 years in Table 26 reveals:

1. At Ibadan, Yoruba children residing in upper class homes surpass those living in underprivileged homes by 3 inches in standing height, and more than 10 pounds in body weight. Similarly large differences are found on comparing "low income" Hindu groups at Delhi and Hyderabad (Table 26) with Hindu children attending elite residential schools in several parts of India. Means for 436 Hindu children measured at these private schools are 51.4 inches and 61.8 pounds.

2. At Hong Kong and Istanbul, Chinese and Turkish children in well-to-do families are taller and heavier than those in economically poor families by at least 1.5 inches and 5 pounds. Dutch children of "prosperous" families exceed United States white children of "poverty" families by over 2 inches in height, and 5 pounds in weight.

3. In the Netherlands, children with parents of high occupational status are taller and heavier than those with parents of low occupational status by slightly less than 1.0 inch, and about 1.5 pounds. These findings are similar to socioeconomic differences obtained in 1963-65 from probability sampling of the United States population, and in 1950 on children residing at Eugene, Oregon.

Table 26 provides the following additional findings:

1. United States white children living in poverty areas are

heavier, and no shorter, than Hong Kong Chinese children of the upper socioeconomic classes.

Table 26

Average Standing Height in Inches and Body Weight in Pounds for Children of Both Sexes Age 9 Years Reared in Privileged and Underprivileged Environments and Measured Since 1960

Group	Number Measured	Mean Weight	Mean Height
Privileged: Upper Socioeconomic Classes			
Filipino, Manila	124	46.9	48.3
Hindu, Delhi	43	46.1	48.8
Hindu, Hyderabad*	246	51.0	49.9
Chinese, Hong Kong	138	54.1	51.0
Baganda, Uganda*	363	61.1	51.3
Turkish, Istanbul	104	64.4	51.7
Yoruba, Ibadan	51	60.3	52.1
Jamaican white, Kingston	117	63.4	52.3
Guatemalan, Guatemala City	49	67.4	52.4
Ghana black, Accra	50	67.1	52.5
Ghana white, Accra	35	67.1	52.9
Dutch, widespread sample	88	65.5	53.9
Underprivileged: Lower Socioeconomic Classes			
Hindu, Hyderabad	175	42.5	46.9
Colombian mestizo, Tenza	159	50.6	46.9
Peruvian mestizo, Lima	52	54.2	47.2
Filipino, Manila	960	44.8	47.3
Colombian, San Jacinto	118	49.4	47.8
Hindu, Delhi	55	45.9	48.0
Chinese, Hong Kong	358	48.3	48.8
Yoruba, Ibadan	57	48.4	48.9
Turkish, Istanbul	110	58.0	50.1
United States white	130	59.4	50.9
United States black	198	61.3	51.7
Dutch, widespread sample	1,139	64.0	53.0

* Girls only, see text.

2. United States black children living in poverty areas are neither shorter nor lighter than Baganda children of Uganda reared under superior conditions of nutritional and health care. Similarly, Dutch children in families of unskilled and semiskilled laborers are no shorter or lighter than Jamaican white children of well-to-do families.

Individual and group differences for size and gain in height and weight between ages 7 and 9 years. From national surveys between 1961 and 1972, averages for standing height at age 7 years are 46.0 inches on Japanese children, 47.3 inches on Czechoslovakian and English children, 47.7 inches on New Zealand and United States children, and 48.6 inches on Dutch children. Averages for body weight are 46.2, 49.3, 50.0, 51.0, 52.3, and 52.5 pounds on children in Japan, England, Czechoslovakia, United States, Netherlands, and New Zealand respectively. For each group, average annual gains from 7 to 8 years are near 2.2 inches in height, and between 5.5 and 6.5 pounds in weight.

Gains from age 7 to age 8 years in the standing height of well-nourished white children are below 1.8 inches for 10%, between 2.1 and 2.5 inches for 40%, and above 2.7 inches for 10%. Corresponding increases in body weight are less than 3.5 pounds for 1 child in 10, 4.9 to 7.3 pounds for 4 in 10, and more than 10.4 pounds for 1 in 10. Gains in height are slightly lower between 8 and 9 years than between 7 and 8 years; in relative terms, height increments average about 4.8% from age 7 to 8 years, and 4.4% from ages 8 to 9 years.

Among 940 Hindu children between ages 7 and 8 years measured in 1965 at rural villages in Palghar Taluk, the shortest and tallest are 39 and 48 inches (99 and 122 cm)—the lightest and heaviest weight 24 and 56 pounds (10.9 and 25.4 kg). Among 1,360 Australian white children between ages 7 and 8 years measured in 1970 at Sydney, 29% are above 48 inches, and 34% above 56 pounds.

Individual differences in height and weight at age 9 years are exhibited in Table 27 by means of 10th and 90th percentiles for each of 12 groups. Comparisons within and between these groups show:

1. Among Belgian children age 9 years living at Courtrai, 10% are shorter than 49 inches, and 10% taller than 55 inches. Other groups in which the lower 10% are no less than 6 inches (15 cm)

shorter than the upper 10% are those of Australian white, Costa Rican, German, Italian, Jamaican black, and United States black and white children.

Table 27

Percentiles for Standing Height in Inches and Body Weight in Pounds at Age 9 Years on Children of Both Sexes Measured Since 1960

Group	Number Measured	Tenth Percentile	Ninetieth Percentile
Standing Height			
Jamaican black	1,625	46.9	53.0
Barbadian black	972	47.9	53.3
French-Canadian, Montreal	563	47.9	53.4
Costa Rican, urban	445	46.7	53.8
Italian, Emilia department	621	48.2	54.3
British, England	1,159	48.7	54.6
Belgian, Courtrai	1,280	49.0	55.0
Australian, Sydney	1,362	48.5	55.2
United States white	1,027	49.1	55.3
Norwegian, Oslo	2,150	49.9	55.6
United States black	485	49.5	55.7
West German, Düsseldorf	7,218	50.0	56.1
Body Weight			
Jamaican black	1,625	45.3	63.4
Barbadian black	972	47.1	65.3
French-Canadian, Montreal	563	48.3	70.8
Costa Rican, urban	445	43.3	71.5
Belgian, Courtrai	1,280	51.3	72.8
British, England	1,159	50.0	76.0
Italian, Emilia department	621	49.6	78.3
Australian, Sydney	1,362	50.6	78.8
Norwegian, Oslo	2,150	54.3	79.3
West German, Düsseldorf	7,218	53.8	80.8
United States white	1,027	51.4	82.1
United States black	485	51.7	85.0

2. Ninety per cent of the children age 9 years on Jamaica Island are less than 53 inches tall. Nearly 50% of their German age peers at Düsseldorf are taller than 53 inches (see Table 24).

3. Among Australian white children age 9 years at Sydney, slightly under 10% weigh 50 pounds or less, and slightly over 10% weigh 78 pounds or more. Other groups in which the lower 10% are at least 28 pounds (12.7 kg) lighter than the upper 10% are those of Costa Rican, Italian, and United States children.

4. Fewer than 10% of children age 9 years residing at Oslo weigh less than 54 pounds. Almost 50% of the children this age living on Barbados Island weigh less than 54 pounds.

Standard deviations for standing height at age 9 years are 2.1 inches (5.4 cm) on 18,300 Japanese children; 2.3 inches on 2,000 Chinese, 1,200 Belgian, and 1,000 Dutch children; 2.4 inches on 12,000 United States white children; 2.5 inches on 1,600 Jamaican black children; 2.6 inches on 21,300 Egyptian children; and about 3.1 inches (near 8 cm) on 7,400 Hindu, and 1,600 Angolan black children. For body weight, standard deviations are 7.5 pounds (3.4 kg) on 1,600 Jamaican black children; 8.2 pounds on 18,300 Japanese, and 7,400 Hindu children; 9 pounds on 21,300 Egyptian, 1,200 Belgian, and 438 Guyanese black children; near 10 pounds on 12,000 United States white children; and between 10 and 12 pounds (4.5 to 5.4 kg) on 14,000 Czechoslovakian, 2,700 Australian, 2,100 Norwegian, and 1,400 Surinam Creole children.

Suggested Readings

Brundtland, G. H., K. Liestöl and L. Walløe **Acta Paediatrica Scandinavica**, 1975, 64, 565-573.

*Goldfeld, A. Y., A. M. Merkova and A. G. Tseimlina **Materials on the physical development of children and adolescents in cities and rural localities of the U. S. S. R.** Leningrad: Meditsina, 1965.

Guaraciaba, M. A. **Jinriu-Gaku Zasshi**, 1967, 75, 1-10.

Hamill, P. V. V., F. E. Johnston and S. Lemeshow **Height and weight of children: socioeconomic status.** (DHEW Publ. No. HSM-73-1601, National Center for Health Statistics, Series 11, No. 119) Washington: Government Printing Office, 1972.

*Low, W. D. **Zeitschrift für Morphologie und Anthropologia**, 1971, 63, 11-45.

*Meredith, H. V. **Monographs of the Society for Research in Child Development,** 1969, 34, No. 1 (Serial No. 125).

New Zealand Department of Health **Physical development of New Zealand school children 1969.** (Health Services Research Unit, Special Report No. 38) Wellington: Government Printer, 1971.

Rona, R. J., and D. G. Altman **Annals of Human Biology,** 1977, 4, 501-523.

CHAPTER X

Size and Form of Head, Trunk, and Limbs in Late Childhood

Head width, head depth, and cephalic index. Averages at age 9 years for the largest transverse and anteroposterior dimensions of the head are assembled in Table 28 from studies on children living in Africa, Asia, Europe, North America, South America, and the islands of Aruba, Palue, and Sicily. Notations on five rows of the table follow:

Mexican, southern Mexico—about equal numbers of these children had Amerind and Spanish progenitors.

Aruba islander—a representative sample of children living on this Netherland Antilles island off the Venezuelan coast.

United States black—children measured at Philadelphia.

Sicilian, two provinces—residents of Messina and Palermo provinces.

Buryat, Siberia—members of Mongolian tribes inhabiting villages in south-central Siberia.

Table 28 shows:

1. In human populations age 9 years, head depth—distance in the midline from between the eyebrows (Glabella) to the back of the head—varies from 6.5 inches (16.5 cm) to 7.2 inches (18.3 cm). On average, the head depth of Uzbek children at Tashkent is smaller than that for children of the Kisi and Nyakyusa tribes in Tanzania by 0.7 inch (1.8 cm), or 10%.

2. Head depth at age 9 years averages 6.8 inches on Chinese and Japanese children, and about 3% higher on Belgian and Dutch children in Europe, and children of the Hutu and Tutsi tribes in Africa.

3. Average width of head at age 9 years is 6.0 inches for Buryat children of south-central Siberia, and 13% less for Hindu children in the Amritsar region of Punjab State.

4. Black children of the Hutu, Kisi, Nyakyusa, Sara, Tutsi, and Yoruba tribes in Africa have heads narrower than Buryat, Chinese, Japanese, and Uzbek children in Asia, and narrower than Belgian, Dutch, Hungarian, and Russian children in Europe.

[89]

Table 28

Averages in Inches for Head Width and Depth (front to back of head) on Children of Both Sexes Age 9 Years Measured Since 1956

Group	Number Measured	Mean Head Depth	Mean Head Width
Hindu, Amritsar area	73	6.9	5.2
Mexican, southern Mexico	66	6.6	5.3
Ethiopian, Gondar region	387	6.7	5.3
Palue islander, Indonesia	92	6.9	5.3
Aruba islander	259	6.9	5.3
Hutu and Tutsi, Rwanda	223	7.0	5.3
Sara, Fort-Archambault	38	6.9	5.4
Italian, Chieti province	520	6.9	5.4
Egyptian, Marsa Matrouh	62	7.0	5.4
Yoruba, Ibadan	62	7.1	5.4
Kisi and Nyakyusa, Tanzania	71	7.2	5.4
Trio and Wajana, Surinam	48	6.7	5.5
United States black	106	7.1	5.5
Sicilian, two provinces	390	7.1	5.6
Chinese, Hong Kong	1,924	6.8	5.6
Dutch, Nijmegen	151	7.0	5.6
Japanese, Hirado and Sasebo	207	6.8	5.7
Belgian, Brussels	847	7.0	5.7
Russian, Moscow	125	6.9	5.8
Hungarian, Körmend	121	6.7	5.9
Polish, rural	72	6.9	5.9
Uzbek, Tashkent	135	6.5	5.9
Buryat, Siberia	154	6.8	6.0

Dependable findings on individual and sex differences in head width are available from a report synthesizing studies prior to 1955 on North American white children. Averages at age 9 years are 5.6 inches (14.2 cm) on 3,900 girls, and 5.7 inches (14.5 cm) on 4,300 boys. The narrowest 10% of head widths among these children are between 5.0 and 5.3 inches, the widest 10% between 6.0 and 6.4 inches.

Comparatively, head width is 5.3 inches or less for 10% of North American white children, and for about 50% of Ethiopian and

Aruban children: it is 6.0 inches or more for 10% of North American white children, and for about 50% of Buryat children.

The cephalic index is lower for black children than oriental children, and intermediate for white children. Head width is between 75% and 79% of head depth on all of the black groups in Table 28, and on the Aruba, Hindu, and Palue groups: It is between 83% and 91% on the Buryat, Chinese, and Japanese groups.

Face height, face width, and face height-width index. Face height is measured from the upper end of the nose (nasion) to below the chin in the midline (menton). Brought together in Table 29 are averages at age 9 years for face height (nasion-menton distance) and face width (bizygomatic distance). It is found:

1. Typically, Asian Buryat children have large—wide and long—faces, and children on Palue Island, East Indies, have small—narrow and short—faces.

2. Face height, on average, is near 4 inches for many groups: examples are Belgian, Dutch, Egyptian, Russian, Sicilian, and United States black groups.

3. Averages for face width are near 4.8 inches on Belgian, Dutch, Hungarian, Russian, Sicilian, and Yoruba children.

Findings from a compilation of studies prior to 1955 on North American white children are:

1. Boys have slightly wider faces than girls. Mean face width at age 9 years is 4.7 inches on 900 boys, and 4.6 inches on 550 girls. The same amount of difference is obtained from the large Belgian study in Table 29.

2. Among North American white children, individuals differ in face width from a little over 4 inches to a little over 5 inches. Specifically, face widths fall between 4.2 and 4.4 inches for 1 child in 10, between 4.5 and 4.9 for 8 children in 10, and between 5.0 and 5.3 inches for the remaining 1 in 10.

3. Increase in average face width during the biennium from age 7 years to age 9 years is slightly less than 0.2 inch (near 0.4 cm). This increase, although small, is twice the increment in head width.

For most groups studied, face height in late childhood averages between 80% and 85% of face width. The Yoruba group (Table 29) has an index of 79, and indices are between 86 and 89 for the Belgian, Egyptian, and Tanzanian groups.

Table 29

Averages in Inches for Face Width (Bizygomatic Diameter) and Face Height (Nasion-Menton Diameter) on Children of Both Sexes Age 9 Years Measured Since 1956

Group	Number Measured	Mean Face Width	Mean Face Height
Palue islander, Indonesia	92	4.6	3.7
Sara, Fort-Archambault	38	4.6	3.7
Mexican, southern Mexico	66	4.8	3.7
Kisi and Nyakyusa, Tanzania	71	4.7	3.8
Yoruba, Ibadan	62	4.8	3.8
Hutu and Tutsi, Rwanda	223	4.5	3.9
Sicilian, Palermo province	190	4.7	3.9
Dutch, Nijmegen	151	4.7	3.9
Hungarian, Körmend	121	4.9	3.9
Egyptian, Marsa Matrouh	62	4.5	4.0
Belgian, Brussels	848	4.7	4.1
United States black	106	4.8	4.1
Russian, Moscow	125	4.9	4.1
Uzbek, Tashkent	135	5.0	4.1
Buryat, Siberia	154	5.2	4.3

Head girth, chest girth, and cephalo-thoracic index. Table 30, constructed from studies since 1960 on head and chest girths in late childhood, reveals:

1. At age 9 years, average chest girth is at or near 24 inches (61 cm) on Chuvash children living in the Soviet Union, and black children residing at Havana and Lourenço Marques. It is larger by about 1.0 inch on Czech and East German children.

2. Compared with Czech and Italian children, Hindu rural children in Bombay State are smaller in average chest girth by about 15% .Mean chest girth on 603 Hindu children measured during 1969-70 at Delhi is the same as for the Hindu rural children.

3. Head girth at age 9 years averages 20.3 inches (51.6 cm) on Slovak children in Europe, Kisi and Nyakyusa children in Africa, and children residing on the islands of Aruba and Sardinia. Averages are slightly higher for Czech and Dutch children.

Table 30

Averages for Head and Chest Girths in Inches on Groups of Children of Both Sexes Age 9 Years Measured Since 1960

Group	Number Measured	Mean Chest Girth	Mean Head Girth
Hindu, rural Bombay	100	21.4	19.4
Chinese, Hong Kong	2,008	------	19.9
Sherpa, Nepal highlands	47	22.7	------
Sara, Fort-Archambault	31	22.9	------
Pedi, Transvaal	40	------	20.2
Sardinian, Sassari	176	23.4	20.3
Cuban black, Havana	155	23.7	------
Aruba islander	258	------	20.3
Moçambique black, Lourenço Marques	395	24.0	------
Chuvash, rural and urban	496	24.2	------
Kisi and Nyakyusa, Tanzania	71	------	20.3
Moldavian, rural and urban	1,002	24.5	------
Slovak, Czechoslovakia	7,000	24.6	20.3
Bulgarian, rural and urban	1,178	24.3	20.4
Uzbek, Tashkent	206	24.6	------
Moçambique white, Lourenço Marques	827	24.8	------
Hungarian, Budapest	546	24.8	------
Russian, Leningrad	779	24.9	------
Dutch, Nijmegen	151	------	20.5
East German, national sample	2,202	25.2	------
Russian, Moscow	407	25.2	------
Czech, Bohemia and Moravia	7,000	25.3	20.5
Italian, Grosseto province	1,045	25.7	21.0

4. Average head girth in late childhood is more than 1.0 inch larger on Italian children of Grosseto province than on Hindu urban children at Delhi and rural children in Bombay State. The mean on 4,900 Hindu children measured between 1956 and 1965 in 12 Indian states is lower than the Grosseto mean by 1.2 inches, or 6%.

5. For Czech, Italian, and Slovak children, mean chest girth at

age 9 years exceeds mean head girth by 4.3 to 4.8 inches. On Hindu children of the Delhi, rural Bombay, and 12-state studies, mean chest girth surpasses mean head girth by 1.5 to 2.4 inches.

6. Relative to head girth, European children are larger in chest girth than Hindu children. Averages for chest girth in percentage of head girth are 115, 119, and 123 on Sardinian, Bulgarian, and Czech children respectively. Corresponding cephalo-thoracic indices are 108 and 112 on Hindu children at Delhi, and in the "all India" survey.

In late childhood, girls are smaller than boys in head and chest girths. For head girth, average differences are 0.2 inch on 4,900 Hindu children, and 0.3 inch on 2,000 Chinese, 10,300 European, and 1,600 North American white children. For chest girth, differences are between 0.4 and 0.8 inch on large samples of East German, Hindu, Russian, and Slovak children.

Among Chinese children at Hong Kong, 10% of individuals age 9 years have head girths smaller than 19.1 inches, and 10% head girths larger than 20.7 inches. There are 10% of Czech children this age below 19.6 inches, and 10% above 21.4 inches.

For Hindu children age 9 years, the "all India" survey shows chest girth is less than 20.6 inches for 1 child in 10, between 20.6 and 25.0 inches for 8 in 10, and more than 25.0 inches for 1 in 10. Hungarian children age 9 years at Budapest vary in chest girth from 21 inches (53 cm) to 33 inches (84 cm).

Using Bulgarian statistics as representative for white children, average increases from age 4 years to age 9 years are 0.6 inch, or 3%, for head girth, and 3.5 inches, or 17%, for chest girth.

Shoulder width, hip width, and shoulder-hip indices. Averages for shoulder and hip widths at age 9 years are assembled in Table 31 on children measured since 1960 in Africa, Asia, Australia, Europe, Central and North America, and the islands of Okinawa, Sardinia, and Taiwan. This table shows:

1. For many groups of white children, average shoulder width (biacromial distance) at age 9 years is near 11 inches (28 cm). Included among these groups are children in Belgium, Bulgaria, Cuba, Estonia, Hungary, Poland, Sardinia, and Western Australia. The average for United States white children is slightly higher.

2. Shoulder widths near 10.5 inches typify Chinese children at Tainan City, Japanese children at Tokyo, Kirghiz children in the

Kirghizian S.S.R., Kisi and Nyakyusa children in Tanzania, and Cakchiquel Maya children in Guatemala. United States white children have wider shoulders than Japanese children at Tokyo by about 6%.

Table 31

Averages in Inches for Biacromial Shoulder Width and Biiliocristal Hip Width at Age 9 Years on Children of Both Sexes Measured Since 1960

Group	Number Measured	Mean Shoulder Width	Mean Hip Width
Sara, Fort-Archambault	31	10.1	7.0
Kisi and Nyakyusa, Tanzania	71	10.4	7.2
Maya Amerind, Guatemala	68	10.2	7.4
Cuban black, Havana	67	10.9	7.5
Yoruba, Ibadan	62	10.9	7.5
South Carolina black	485	------	7.6
Taiwanese, Tainan City	100	10.6	7.6
Pedi, Transvaal	40	11.5	7.6
Japanese, Tokyo	234	10.7	7.7
Okinawan, Nakijin	22	11.0	7.7
Cuban white, Havana	105	10.8	7.9
Kirghiz, Osh and Kirov areas	201	10.6	8.0
Sardinian, Sassari	176	10.8	8.0
Polish, Nowy Targ region	202	10.9	8.0
United States white	1,027	11.4	8.0
Polish, Lublin	365	10.9	8.1
Belgian, Brussels	848	11.1	8.1
Australian white, Perth	957	11.2	8.1
Guatemalan Spanish	49	11.3	8.1
Hungarian, Budapest	546	11.2	8.2
Bulgarian, rural and urban	1,180	10.9	8.3
Russian, Moscow	407	11.0	8.3
Estonian, Tallinn	418	11.2	8.3

3. Hip width (distance between the iliac crests) at age 9 years is narrower for black than white children. Means are between 7.0 and 7.6 inches on black children of Chad, Cuba, Nigeria, Tanzania, and the United States; they are between 8.0 and 8.3 inches

on white children in Australia, Belgium, Bulgaria, Estonia, Hungary, Poland, the Soviet Union, and the United States. Compared with black children at Havana and Ibadan, white children at Tallinn and Moscow have hips averaging 10% wider.

4. For Japanese and Taiwan Chinese children age 9 years, mean shoulder width exceeds mean hip width by 3.0 inches. Differences are near 3.3 inches (8.5 cm) on black children, and near 2.9 inches (7.5 cm) on several white groups.

During late childhood, boys and girls are about alike in widths of the trunk. At ages 8 and 9 years, in both shoulder width and hip width, averages for the two sexes differ by less than 0.2 inch on Bulgarian children, Australian children at Perth, Hungarian children at Budapest, Japanese children at Tokyo, United States black children at Columbia and Philadelphia, and United States white children.

Among 200 well-cared-for United States white children age 7 years, 20% have shoulder widths either narrower than 10.1 inches, or wider than 11.3 inches. Increases between ages 7 and 8 years are less than 0.3 inch for 10%, 0.3 to 0.6 inch for 80%, and more than 0.6 inch for 10%. Shoulder widths at age 9 years on 1,300 Australian children at Sydney vary from 8.3 to 10.8 inches for 10%, and from 12.4 to 15.7 inches for 10%.

Twenty per cent of 200 United States white children age 7 years have hip widths either narrower than 7.3 inches, or wider than 8.3 inches. Increases in this width between ages 7 and 8 years are less than 0.2 inch for 10%, 0.2 to 0.4 inch for 80%, and more than 0.4 inch for 10%. Hip widths at age 9 years on 485 United States black children are less than 7.0 inches for 1 child in 10, and more than 8.3 inches for 1 in 10: on 1,027 United States white children, hip widths exceed 8.9 inches for 1 child in 10.

Mean hip width at age 9 years relative to mean shoulder width is near 69% for black children in Cuba, Nigeria, and Tanzania; near 72% for Chinese and Japanese children; and near 74% for Belgian, Estonian, Hungarian, Polish, and Sardinian children. These values register the relative narrowness of hip width in different groups. Reciprocal indices showing the relative predominance of shoulder width are near 144 for black children, 139 for Chinese and Japanese children, and 136 for Bulgarian, Estonian, Hungarian, Polish, and Sardinian children.

Sitting height, lower limb height, and skelic index. Compiled in

Table 32 are averages for sitting height and lower limb height on children age 9 years inhabiting numerous regions of the earth. Findings are as follows:

Table 32

Averages in Inches for Sitting Height and Lower Limb Height on Children of Both Sexes Age 9 Years Measured Since 1960

	Number Measured	Mean Sitting Height	Mean Lower Limb Height
Bundi, New Guinea highlands	90	23.8	19.5
Guatemalan Amerind	68	25.2	20.6
Quechua, Peru highlands	78	26.2	20.7
Filipino, Manila*	637	25.5	21.7
Brazilian Japanese, Bauru	275	26.9	21.9
Trio and Wajana, Surinam	48	24.3	22.3
Japanese, 1970 survey	18,346	27.7	22.5
Chinese, Hong Kong	2,098	26.9	22.7
Bulgarian, rural and urban	1,173	27.2	23.3
Hindu, Delhi	603	24.8	23.6
Italian, Chieti province	520	26.6	23.6
Tanzanian black	173	25.4	23.8
Aruba islander	258	27.6	23.8
New Zealand Maori	400	27.8	24.0
Sara, Fort-Archambault	31	25.5	24.1
Australian white, Perth	957	27.6	24.1
New Zealand white	1,816	27.7	24.4
Hungarian, Budapest	546	27.7	24.5
United States white	1,027	27.7	24.5
North German, Kiel	106	28.1	24.9
Cuban black, Havana	155	26.2	25.0
South Carolina black	485	27.6	25.0
Australian aborigine	20	24.9	26.1

* Girls only.

1. Compared with Maori children in New Zealand, Bundi children in New Guinea are shorter in mean sitting height at age 9 years by 4 inches, or 14%. The body stem of Maori children typically exceeds that of Tanzanian black and Hindu children by more than 2 inches, or about 9%.

2. Averages for sitting height at age 9 years are near 27.5 inches (70 cm) on Aruba islanders, Australian white children at Perth, Hungarian children at Budapest, Japanese children, New Zealand white children, South Carolina black children, and United States white children living in the 48 contiguous states.

3. Averages for lower limb length at age 9 years are near 21 inches (53 cm) for Quechua children in Peru; 23 inches for Bulgarian children, and Chinese children at Hong Kong; and 24 inches (61 cm) for Australian white, New Zealand Maori, Tanzanian black, and Aruba island children. Black children age 9 years in Cuba and South Carolina, on average, have longer lower limbs than their Bundi age peers in New Guinea by fully 5 inches, or 26%.

4. In late childhood, average sitting height exceeds average lower limb height by about 5 inches on native Japanese children, and by no more than 2 inches on black children in Chad, Cuba, and Tanzania. Lower limb height is between 80% and 85% of sitting height on Bundi, Chinese, and Japanese children; near 88% on Hungarian, Italian, New Zealand white, and United States white children; between 90% and 95% on black children of Chad, Cuba, Tanzania, and the United States; and near 105% on Australian aboriginal children.

Table 32 includes large samples of Australian, Bulgarian, Chinese, Japanese, and United States white children. Subgrouped by sex, these samples show that at age 9 years boys surpass girls by 0.2 to 0.3 inch in mean sitting height, and the sexes are almost alike (differ no more than 0.1 inch) in mean lower limb height.

Among United States white children age 7 years, 20% have sitting heights either below 25 inches or above 28 inches. Between ages 7 and 8 years, sitting heights of white, healthy, well nourished individuals increase less than 0.6 inch for 1 child in 10, 0.6 to 1.2 inches for 8 in 10, and more than 1.2 inches for 1 in 10. At age 9 years, sitting heights of United States white children are below 26 inches for 1 child in 10, and above 29 inches for 1 in 10.

At age 7 years, 20% of United States white children have lower limb heights either under 20.7 inches, or over 23.7 inches. Increments between ages 7 and 8 years are less than 1.1 inches for 1 child in 10, and more than 1.7 inches for 1 in 10. Among South Carolina black children age 9 years (Table 32), lower limb heights are between 21.7 and 23.1 inches for the shortest 10%, and between 26.8 and 30.0 inches for the tallest 10%.

Individual differences in the skelic index at age 9 years extend from 76 to 98 on 117 South Carolina white girls, and from 78 to 105 on 277 South Carolina black girls. Indices are at or above 81 for 50% of the white girls, and 90% of the black girls.

Upper limb length, arm girth, leg girth, and limb indices. From studies in the 1960's on Belgian and Bulgarian children, means for upper limb length at age 7 years are 20.2 inches (51.3 cm) on boys, and 0.3 inch lower on girls. Means at this age from measures taken between 1938 and 1949 on United States boys of Italian, Finnish, and northwest European ancestry are 20.0, 20.4, and 20.8 inches respectively. Among United States white children of both sexes age 7 years, 10% have upper limb lengths between 18.5 and 19.5 inches, and 10% between 21.7 and 23.2 inches. Increases in upper limb length between ages 7 and 8 years are less than 0.8 inch for 1 child in 10, 0.8 to 1.3 inches for 8 in 10, and more than 1.3 inches for 1 in 10. At age 9 years, means for the sexes together are 22.1 inches (56 cm) on 848 Belgian and 1,180 Bulgarian children, 22.5 inches on 419 Estonian children, 22.6 inches on 509 United States white children at Iowa City and Philadephia, and 23.0 inches on 102 United States black children at Philadephia.

Statistics are given in Table 33 for arm girth and calf girth at age 9 years. This table shows:

1. Compared with average calf girth of Pedi children at age 9 years living on the Sekhukhune Bantu reserve, average calf girth of South Carolina black children is greater by about 2 inches, or 25%. Averages are intermediate for black groups in Chad and Tanzania.

2. Calf girth at age 9 years, on average, is near 10.5 inches for United States white children, English children in London County, and United States black children in South Carolina. The average is slightly higher for children of predominantly Spanish ancestry attending private schools in San Juan.

3. Average arm girth at age 9 years is near 6.7 inches on black children in rural Guyana, Hindu children of wealthy families at Hyderabad, and Yoruba children of poor families at Ibadan. Means are higher (between 7.5 and 8.0 inches) on English children in London County, Belgian children at Brussels, United States black children in South Carolina, and United States white children.

4. Within a given population, arm girth in late childhood is

larger for well-nourished children than underprivileged children. At Ibadan, average arm girth is about 0.8 inch greater for 51 children of well-to-do families than 58 children of poor families. At Hyderabad, 246 girls in upper-class homes exceed 55 girls in lower-class homes by about 0.4 inch. Since most of these samples are small, the direction of findings is more dependable than the amount of difference.

Table 33

Averages in Inches for Arm Girth and Calf Girth on Children of Both Sexes Age 9 Years Measured Since 1960

Group	Number Measured	Mean Calf Girth	Mean Arm Girth
Hindu, Delhi	603	8.6	6.2
Guyanese Hindu	236	----	6.3
Pedi, Transvaal	40	8.4	6.3
Egyptian, Khartoum	165	----	6.3
Zapotec, Mexico	40	----	6.5
Sara, Fort-Archambault	31	9.5	6.6
Maya Amerind, Guatemala	68	----	6.6
Hindu, wealthy at Hyderabad*	246	----	6.7
Yoruba, poor at Ibadan	58	----	6.7
Guyanese black, rural	181	----	6.7
Kisi and Nyakyusa, Tanzania	71	9.3	6.9
Trio and Wajana, Surinam	35	9.4	7.0
African black, Dar es Salaam	65	----	7.0
Bulgarian, rural and urban	1,180	----	7.2
English, London County**	2,325	10.6	7.5
Yoruba, wealthy at Ibadan	51	----	7.5
Belgian, Brussels	848	----	7.6
United States white	1,027	10.4	7.7
United States black	485	10.6	7.9
Guatemalan Spanish	49	----	8.0
Puerto Rican Spanish*	174	10.9	----

* Girls only.
** These children were measured in 1959.

Girls and boys are similar at late childhood ages in average girth of both arm and calf. From measures on 6,700 English children ages 7, 8, and 9 years, differences between separate means for the sexes do not exceed 0.1 inch. The same similarity is found from measures at age 9 years on 1,027 United States white girls and boys, and 485 South Carolina black girls and boys.

Among 2,100 English children age 7 years, arm girths are smaller than 6.2 inches (15.7 cm) for 10%, and larger than 7.8 inches (19.8 cm) for 10%. Corresponding values are 6.5 and 8.1 inches on 218 United States white private school children. Increases among this group between ages 7 and 8 years are less than 0.1 inch for 10%, 0.1 to 0.6 inch for 80%, and more than 0.6 inch for 10%. At age 9 years, arm girths on 1,027 United States white children are below 6.7 inches for 10%, and above 9.0 inches for 10%. Comparable values on 485 United States black children are 6.8 and 9.5 inches.

Drawing from the same studies, calf girths at age 7 years on children of London County are smaller than 8.8 inches (22.5 cm) for 10%, and larger than 10.7 inches (27.2 cm) for 10%. Corresponding values on Iowa white private school children are 8.9 and 11.0 inches. Increments between ages 7 and 8 years among well-nourished white children are under 0.2 inch for 1 child in 10, 0.2 to 0.7 inch for 8 in 10, and over 0.7 inch for 1 in 10. Calf girths at age 9 years on United States white children are below 9.3 inches for 10%, and above 11.6 inch for 10%. Among 174 Puerto Rican private school children age 9 years, calf girths are between 9.1 and 9.7 inches for the smallest 10%, and between 12.4 and 14.8 inches for the largest 10%.

For United States white private school children age 7 years, arm girth averages 36% of upper limb length, and calf girth averages 45% of lower limb height. Indices on 20% of these children are either below 32 or above 39 for the upper limb, and below 41 or above 49 for the lower limb. At age 9 years, averages for the lower limb index are 42 on both 1,027 United States white children, and 485 South Carolina black children. Indices on the South Carolina black children are between 34 and 37 for 1 child in 10, and between 48 and 57 for 1 in 10.

From age 4 years to age 9 years, average increases for arm and calf girth of white children are near 1.0 and 1.8 inches respectively. Throughout this period, both upper and lower limbs become more slender: average decreases are from 40 to 35 for arm girth in percentage of upper limb length, and from 50 to 44 for calf girth in percentage of lower limb height.

Standard deviations on white children age 9 years are near 0.2 inch for head width and face width; 0.6 inch for head girth, shoulder width, and hip width; 0.9 inch for arm girth and calf girth; 1.2 inches for sitting height and upper limb length; 1.4 inches for chest girth and lower limb height; and 4 percentage points for the limb indices. Compared with these values on white children, standard deviations on Hindu children are similar for head girth, hip width, calf girth, and sitting height, but lower for arm girth (0.5 inch).

Suggested Readings

Banik, N. D. D., S. Nayar, R. Krishna, L. Raj and A. D. Taskar **Indian Journal of Pediatrics,** 1970, 37, 453-459.

Correnti, V. **Rivista di Antropologia,** 1969, 55, Supplement, 1-210.

Krogman, W. M. **Monographs of the Society for Research in Child Development,** 1970, 35, No. 3 (Serial No. 136).

Malina, R. M., P. V. V. Hamill and S. Lemeshow **Body dimensions and proportions, white and negro children 6-11 years.** (DHEW Publ. No. HRA-75-1625, National Center for Health Statistics, Series 11, No. 143) Washington: Government Printing Office, 1974.

Meredith, H. V. **Growth,** 1971, 35, 233-251.

Meredith, H. V., and E. M. Meredith **Child Development,** 1953, 24, 83-102.

*Spurgeon, J. H., E. M. Meredith and H. V. Meredith **Annals of Human Biology,** 1978, 5, No. 3.

*Yanev, B. (Ed.) **Physical development and fitness of the Bulgarian people from birth to the age of twenty-six.** Sofia: Bulgarian Academy of Sciences Press, 1965.

CHAPTER XI

Change During the Last Century in Body Size of Children

Few studies on the body size of children were made prior to 1870. In the decades since, findings have accumulated showing children in many parts of the world have increased from generation to generation in body weight, body length or height, and certain other external dimensions. The term used to designate this phenomenon is secular change.

Secular change in body size at ages 1 year and 4 years. From data collected during 1918-19 on United States infants age 1 year, means for body weight and length are 19.8 pounds and 28.8 inches for 6,476 white infants and, for 182 black infants, 19.0 pounds and 28.1 inches. These groups were obtained without regard to parental socioeconomic status. A study at the same age, using measures taken during 1969 on United States infants living in poverty areas, yields means higher by 1.3 pounds and 0.6 inch for 213 white infants, and by 1.5 pounds and 1.1 inch for 533 black infants. Alternative pairings of weight and length averages show infants of middle class families measured since 1960 (Table 11) exceed those in the 1918-19 study by 2.2 pounds and 0.8 inch for white infants, and by 3.3 pounds and 1.4 inches for black infants.

At age 4 years, comparisons are available for Belgian and United States white groups measured at times separated by about 130 and 90 years respectively. On 455 Belgian children measured during 1960-61 at Brussels, means are 39.6 inches for height and 36.2 pounds for weight. These values surpass means on Belgian children of "all social classes" studied at Brussels during the early 1830's by no less than 3 inches in height, and 6 pounds in weight. From data on 201 United States white children collected during 1971-72 in 23 states and the District of Columbia, average height (40.2 inches) exceeds by 2.1 inches that on 110 United States white children measured in 1881 at Milwaukee.

Other comparisons at age 4 years are presented in Table 34. Since the time span is not the same for all rows of the table, each difference between means (column 5) is transposed to the amount of increase per decade (column 6).

[103]

Table 34

Averages for Height in Inches and Weight in Pounds on Children of Both Sexes, Age 4 Years, Measured Since 1890 at Times 50 to 70 Years Apart

Early Study Number	Mean	Later Study Number	Mean	Later Mean Higher by:	Increase per Decade
English: Urban Centers, 1909-10; London, 1959*					
20,816**	38.3	286	41.5	3.2	.65
Same	34.0	Same	37.3	3.3	.67
French: Paris, 1905 and 1957-63					
265	36.4	296	39.6	3.2	.58
Same	29.6	Same	34.6	5.0	.91
Australian White: Sydney, 1901-07 and 1970-72*					
78	38.6	2,258	41.5	2.9	.43
Same	36.4	2,252	39.8	3.4	.51
German: Halle, 1891-1900; Einbeck and Kiel, 1952					
92†	37.6	139	40.0	2.4	.43
United States White: Gary, Ind., 1918; Ypsilanti, Mich., 1968-73					
307	38.0	114	39.7	1.7	.33
Same	32.3	Same	36.6	4.3	.83
United States Black: General Samples, 1918-19 and 1968-72					
385	39.1	171	40.5	1.4	.27
Same	33.6	Same	36.2	2.6	.51

* Age 4.5 years.

** For each pair of rows, the upper row pertains to height and the lower row to weight.

† Height only.

It can be concluded:

1. During the last century, at ages from late infancy through early childhood, overall body size has increased. Increases in average height and weight are found consistently, although the amounts of increase vary in different regional and racial groups.

2. At age 4 years, among the several groups spanning 50 to 55 years, the smallest increases are near 1.5 inches in height, and 2.5 pounds in weight.

3. Taking the studies assembled at age 4 years as a whole, average increases per decade are near 0.4 inch and 0.7 pound for standing height and body weight respectively.

Secular change in height and weight at age 9 years. Contained in Tables 35 and 36 are 18 separate comparisons for secular change in standing height at age 9 years. Table 35 displays changes during periods of 50 to 80 years, and Table 36 shows changes in periods of 30 to 45 years. From these tables, it is found:

1. Secular increase in standing height is common to populations of children age 9 years in Australia, Asia, Europe, and North America.

2. The 6 groups studied at times 65 to 75 years apart increased in height between 3 and 5 inches, averaging 3.8 inches. These are the groups of Australian white, Czech, Japanese, Russian, United States black, and United States white children in Table 35. Average increase for the 6 groups studied between 30 and 45 years apart (Table 36) is 2.3 inches.

3. For white children age 9 years, one-half inch per decade can be considered to typify average increase in standing height during the last several decades. The numerous comparisons on children of European ancestry provide a strong base for this generalization. When the entire 18 populations in Tables 35 and 36 are combined, average increase in height per decade is between 0.5 and 0.6 inch.

Secular changes for body weight at age 9 years are listed in Tables 36 and 37. These tables show:

1. Increase in recent decades for each of the 18 groups.

2. Increases varying from 5 to 13 pounds for the groups studied at times separated by 50 to 80 years.

3. A composite increase near 1.5 pounds in body weight per decade.

For children of a given population, increase in average standing height could result from either individuals in one portion of the height distribution becoming taller, or individuals in all parts of the distribution getting taller. At several ages in middle and late childhood, it has been discovered that during recent decades

Table 35

Averages for Standing Height in Inches on Children of Both Sexes, Age 9 Years, Measured Since 1880 at Times from 50 to 80 Years Apart

Early Study Number	Mean	Later Study Number	Mean	Later Mean Higher by:	Increase per Decade
\multicolumn{6}{c}{Norwegian: Oslo, 1920 and 1970}					
3,100*	49.5	2,150	52.9	3.4	.68
Czech: Bohemian and Moravian Surveys, 1894-95 and 1968-70					
8,000*	47.9	3,400*	52.6	4.7	.64
English: National Surveys, 1909-10 and 1972					
18,854	47.8	1,159	51.7	3.9	.63
Russian: Moscow, 1887 and 1961-62					
425	46.9	408	51.5	4.6	.62
Japanese: National Surveys, 1900 and 1970					
10,000*	46.2	18,349	50.3	4.1	.59
East German: Saalfeld District, 1889; East Germany, 1967-68					
1,160	47.9	2,202	52.2	4.3	.55
French: Paris, 1905 and 1962-68					
326	48.1	210	51.2	3.1	.52
Swedish: Stockholm, 1883 and 1938-39					
198	49.9	199	52.7	2.8	.51
United States White: St. Louis, 1892; National Sample, 1963-65					
4,327	48.8	1,027	52.2	3.4	.47
United States Black: Washington, D. C., 1896-98 and 1963-65					
512	49.2	116	52.3	3.1	.46
Australian White: Sydney, 1901-07 and 1970					
2,883	48.8	988	51.8	3.0	.45
Canadian White: Toronto, 1892; London, 1967-69					
1,773	48.6	500	51.5	2.9	.38

* About this number measured

height and weight distributions have moved in their entirety toward greater tallness and heaviness. Examples of these shifts are accessible from studies on United States black and white children. To wit: standing heights at age 9 years are below 49 inches for 50% of 4,300 white children measured in 1892, and 10% of 600 white children measured during 1963-65. Heights are above 52 inches for 10% of the 1892 group, and 50% of the 1963-65 group;

Table 36

Averages for Height in Inches and Weight in Pounds on Children of Both Sexes, Age 9 Years, Measured Since 1920 at Times 30 to 45 Years Apart

Early Study		Later Study		Later Mean	Increase
Number	Mean	Number	Mean	Higher by:	per Decade
Kazak: Alma-Ata, 1930-33 and 1960-62*					
67**	47.4	108	50.4	3.0	1.0
Same	51.8	Same	59.1	7.3	2.5
Russian: Syktyvkar, 1927 and 1967-68					
400†	48.1	580	51.3	3.2	0.8
Same	53.6	Same	59.5	5.9	1.5
East German: Jena, 1921 and 1954-55					
380†	48.7	230	51.2	2.5	0.8
Same	55.9	Same	59.1	3.2	1.0
Sardinian: Sassari Province, 1927-30 and 1965					
280†	47.8	200†	49.7	1.9	0.5
Same	50.7	Same	58.9	8.2	2.3
United States Black: San Antonio, 1929-31; Columbia, 1974-77					
286	50.8	485	52.6	1.8	0.4
Same	56.6	Same	65.7	9.1	2.0
Canadian White: Toronto, 1923; Halifax, 1968-69					
7,882	50.1	574	51.6	1.5	0.3
Same	58.0	Same	62.1	4.1	0.9

* Boys only.
** For each pair of rows, the upper row pertains to height and the lower row to weight.
† About this number measured.

Table 37

Averages for Body Weight in Pounds on Children of Both Sexes, Age 9 Years, Measured Since 1880 at Times 50 to 80 Years Apart

Early Study Number	Mean	Later Study Number	Mean	Later Mean Higher by:	Increase per Decade
Norwegian: Oslo, 1920 and 1970					
3,100*	55.0	2,150	65.7	10.7	2.1
Russian: Moscow, 1887 and 1961-62					
425	49.3	408	62.4	13.1	1.8
Czech: Bohemian and Moravian surveys, 1894-95 and 1968-70					
8,000*	52.8	3,400*	65.5	12.7	1.7
French: Paris, 1905 and 1962-68					
326	49.6	210	59.7	10.1	1.7
United States Black: Atlanta, 1925-26; Columbia, 1974-77					
624	58.2	485	65.6	7.4	1.5
English: National Surveys, 1909-10 and 1972					
18,854	51.9	1,159	60.5	8.6	1.4
Japanese: National Surveys, 1900 and 1970					
10,000*	47.4	18,349	57.4	10.0	1.4
East German: Saalfeld District, 1889; East Germany, 1967-68					
1,160	52.3	2,202	62.8	10.5	1.3
Australian White: Sydney, 1901-07 and 1970					
2,883	55.1	988	63.7	8.6	1.3
United States White: St. Louis, 1892; National Sample, 1963-65					
4,243	56.4	1,027	65.2	8.8	1.2
United States Black: Washington, D. C., 1896-98 and 1963-65					
512	56.9	116	64.6	7.7	1.1
Swedish: Stockholm, 1883 and 1938-39					
198	59.0	199	63.9	4.9	0.9

* About this number measured.

weights are 56 pounds or less for 60% of the 1892 group, and 25% of the 1963-65 group; and weights are 70 pounds or more for 5% and 25% of the respective groups. On United States black children age 9 years, heights are less than 49 inches for 50% of 500 individuals measured during 1896-98, and 8% of 485 measured during 1974-77. Weights are above 57 pounds for 50% of the 1896-98 group, and 70% of the 1974-77 group.

Secular change at age 9 years in sitting height, lower limb height, and chest girth. In Table 38 are statistics relating to secular change in sitting height, lower limb height, and chest girth. Findings from this table are as follows:

1. Among black, oriental, and white populations, secular increase has occurred within the last 100 years in the body stem and lower limb components of standing height.

2. Averages for sitting height of United States black and white children age 9 years have increased at least 1.5 inches during the last 75 years. There have been similar amounts of increase in the averages for lower limb height.

3. During the last 30 years, on average, Japanese children age 9 years have increased in sitting height about 0.3 inch per decade, and in lower limb height about 0.5 inch per decade.

4. There has been secular increase in chest girth during recent decades. For various intervals since 1887, the three groups of children age 9 years represented in the lower part of Table 38 increased in average chest girth about 0.2 inch per decade.

Averages for face width, arm girth, and calf girth on United States white children age 9 years measured in the 1960's exceed corresponding averages obtained in the 1930's by 0.1, 0.3, and 0.2 inch respectively. Similar slight differences are found at other childhood ages.

Secular change beyond childhood. The progressive secular increases in human body size described from infancy through chilhood continue into adolescence. Findings on standing height suffice to illustrate this. Averages for height of females age 12 years, and males age 14 years, have increased during the last 70 years by 5 to 6 inches on Japanese youths, and by 4 to 6 inches in United States black youths, and white youths in Australia, Belgium, Great Britain, the Netherlands, Norway, Poland, the Soviet Union, Sweden, and

the United States. Australian females age 12 years were taller in
1970 than 1904 by 4 inches, or 7%, and Japanese females this age
were taller in 1970 than 1900 by 6 inches, or 11%.

Table 38

Averages in Inches for Sitting Height, Lower Limb Height, and Chest Girth at Age 9 Years on Children of Both Sexes Measured Since 1885 at Times Separated by 30 to 80 Years

Early Study		Later Study		Later Mean	Increase
Number	Mean	Number	Mean	Higher by:	per Decade
Sitting Height					
United States Black: Washington, D. C., 1896-98; Columbia, 1947-77					
512	25.9	485	27.6	1.7	.22
United States White: St. Louis, 1892; National Sample, 1963-65					
4,329	26.1	1,027	27.7	1.6	.22
Japanese: National Surveys, 1937 and 1970					
10,000*	26.6	18,349	27.7	1.1	.33
Lower Limb Height					
United States Black: Washington, D. C., 1896-98; Columbia, 1974-77					
512	23.3	485	25.0	1.7	.22
United States White: St. Louis, 1892; National Sample, 1963-65					
4,329	22.7	1,027	24.5	1.8	.25
Japanese: National Surveys, 1937 and 1970					
10,000*	21.0	18,349	22.6	1.6	.48
Chest Girth					
Kazak: Alma-Ata, 1930-33 and 1960-62					
67†	23.9	108	24.5	0.6	.21
Japanese: National Surveys, 1900 and 1970					
10,000*	22.8	18,349	24.3	1.5	.21
Russian: Moscow, 1887 and 1961-62					
430	23.1	408	25.2	2.1	.29

* About this number measured.
† Boys only.

Secular changes at adult ages have been much smaller than at mid-adolescent ages. For instance: on 52,000 United States men between ages 20 and 25 years measured during 1863-64, mean standing height is 67.8 inches; on a comparable group of 625 men measured 108 years later, the mean is higher by 2.0 inches—yielding an increase of 3%, or 0.2 inch per decade. A recent integrative report on secular change concludes that during the last century increases in height per decade have averaged "one-half inch in late childhood, three-fourths inch at mid-adolescence, and one-fourth inch in early adulthood."

What has caused secular increase in recent decades? Many suggestions have been made on possible causes of secular change. These include: reduction in frequency and severity of illness, increase in nutritional content of food, reduction in child labor, greater participation in sports, improvement in personal and community hygiene, decrease in family size, effect of increasing urbanization, population change due to migrations and wars, effect of selective mating, change in world temperature and humidity, expression of hybrid vigor, and evolutionary modification of the pituitary growth hormone.

Probably there are multiple determinants. Combinations of two or more variables may account for secular change at different ages, in different populations, and during different decades. One biologist has noted that secular increase began before the era of scientific medicine; another has observed that secular increase began before the introduction of vitamin supplements, or enriched and fortified foods.

Although many of the attempts to identify causes have been thought provoking, they should be recognized as not extending beyond plausible conjecture. To date, quoting the summarization of one writer, "nobody knows for certain why the secular trend has occurred."

Suggested Readings

Knott, V. B., and H. V. Meredith **Human Biology,** 1963, 35, 507-513.

*Meredith, H. V. Change in the stature and body weight of North American boys during the last 80 years, in L. P. Lipsitt and C. C. Spiker (Eds.), **Advances in child development and behavior.** New York: Academic Press, 1963 (Volume 1, 69-114).

*Meredith, H. V. American Journal of Physical Anthropology, 1976, 44, 315-326.

*Meredith, H. V. Growth, 1978, 42, 37-41.

Suchý, J. Review of Czechoslovak Medicine, 1972, 18, 18-27.

Tanner, J. M. Tijdschrift Voor Sociale Geneeskunde, 1966, 44, 524-539.

Wieringen, J. C. van. Seculaire groeiverschuiving: Lengte en gewicht surveys 1964-1966 in Nederland in historisch perspectief. Leiden: Netherlands Instituut voor Praeventieve Geneeskunde, 1972.

Chapter XII

Dental Changes During Middle and Late Childhood

Shedding of primary teeth. A British book published in 1834 gives "approximate" average ages for shedding of the primary incisor teeth. To wit: between ages 7 and 8 years, "first the two central incisors of the lower jaw fall out," then "the two central incisors of the upper jaw fall out"; and between 8 and 9 years, there is "shedding of the lower lateral, then the upper lateral incisors." Later investigators have confirmed this sequence, but found earlier timing. In a 1954 report on "several thousand" English children studied at London, it is stated that 95% of those age 7 years had shed their primary lower central incisors.

On Bantu children of the Digo tribe examined at coastal villages in Kenya, average ages of shedding primary teeth are 6.0 years for the lower central incisors, 6.5 years for the upper central incisors, 7.0 years for the lower lateral incisors, and 7.9 years for the upper lateral incisors. On United States white children at New York, findings for mean age at which children of wealthy families discard primary teeth are 6.3 years for the lower central incisors, 7.0 years for the upper central incisors, 7.2 years for the lower lateral incisors, and 7.9 years for the upper lateral incisors. Statistics comparing this sample with a New York white sample of poor children show: shedding is slightly later among the poor than the wealthy, and slightly earlier for girls than boys. Taking the upper lateral incisors to illustrate these findings, means are 7.6, 8.2, 8.3, and 8.5 years for the upper class girls, upper class boys, lower class girls, and lower class boys respectively.

Few children shed a primary tooth before age 5 years, and few fail to shed all primary incisors (4 central and 4 lateral) by age 9 years. There are rare instances of congenitally absent permanent lower central incisors, with retention of the primary lower central incisors into adulthood.

Oral emergence of permanent dentition in perspective. At the time of oral emergence, teeth have had a long prior development, and are not fully formed. Years of gradual tooth formation both

[113]

precede and follow the age when a given permanent tooth breaks through the gum and becomes visible on oral examination. Formation of enamel and dentin usually commences during the first year after birth in permanent first molar, incisor, and canine teeth; between ages 1 and 3 years in permanent premolar and second molar teeth; and during late childhood in permanent third molar teeth. The permanent first molar will serve to illustrate the extensive course of formation. For this tooth, initial enamel and dentin cells appear near the time of birth, half the crown is completed by age 2 years, completion of the crown is reached by ages 3 to 4 years, half the root is completed between ages 5 and 6 years, and completion of the root occurs at varying ages from 7 to 12 years.

Oral emergence of a tooth is fitly recognized as one stage in a process of movement that begins earlier and continues later. Several months elapse between the age a tooth pierces the alveolar bone and the age it pierces the gingival tissues: the average intervals found in one study are 8 months for the permanent lower first premolar, and 11 months for the permanent lower canine. It takes another several months for a tooth to migrate from gingival emergence to contact with teeth of the opposite jaw. For example, the permanent lower incisor and canine teeth take 2 to 7 months to travel 50% of this distance, 3 to 15 months to travel 70% of the distance, and 15 to 48 months to fully attain the level of occlusion.

Oral emergence of permanent first molar and incisor teeth. Almost always oral emergence of the permanent dentition begins with perforation of the mandibular (lower jaw) gingiva by a first molar or central incisor tooth. (The permanent first molars emerge behind the primary dentition, immediately posterior to the primary second molars, while the permanent central incisors replace the primary central incisors.)

Presented in Table 39 are averages for age of oral emergence of the permanent first molar and central incisor teeth. Findings from this table on children studied since 1940 are as follows:

1. For the human groups represented, taken as a whole, the typical ages at which permanent teeth pierce the oral gingivae are 5.8, 6.2, 6.2, and 7.0 years for the lower first molars, upper first molars, lower central incisors, and upper central incisors respectively. On average, the earliest permanent teeth to emerge are the lower first molars; these are succeeded in about 5 months by emer-

gence of the upper molars and lower central incisors, and in about
15 months by emergence of the upper central incisors.

Table 39

**Average Age in Years at Oral Emergence of the Permanent First
Molar and Central Incisor Teeth in Children of
Both Sexes Studied Since 1940**

Group	First Molars Lower	Upper	Central Incisors Lower	Upper
Siassi islander, near New Britain	5.4	5.5	6.0	6.4
Bougainville islander	5.8	5.8	6.0	6.5
Ghana black, Sunyani region	5.2	5.5	5.7	6.6
Kaiapit, New Guinea	5.4	5.7	6.3	6.6
Digo Bantu, Kenya	5.3	5.4	5.8	6.7
Polish, Warsaw	5.8	6.2	6.2	6.7
English, Birmingham	6.0	6.0	5.9	6.8
United States black, Michigan	6.1	6.3	5.9	6.8
Bundi, New Guinea	5.6	5.7	6.5	7.0
Brazilian Japanese, São Paulo	5.9	6.2	6.3	7.0
Australian white, New South Wales	6.0	6.2	6.1	7.0
United States white, Iowa City	6.1	6.2	6.0	7.0
New Zealand white, Dunedin	6.4	6.4	6.3	7.0
Australian aboriginal	6.1	6.2	6.6	7.1
Mandinka, Gambian villages	5.6	5.9	6.2	7.2
Polish, rural districts	5.6	6.5	6.3	7.2
United States white, Michigan	6.2	6.3	6.3	7.2
Hong Kong Chinese	6.0	6.3	6.2	7.3
Japanese, Nagasaki	5.6	6.3	6.5	7.3
Taiwan Chinese	5.9	6.3	6.4	7.3
Sherpa, Nepal highlands	6.1	6.5	7.5	7.3
Pima Amerind, Arizona	5.7	5.9	6.2	7.6

2. Averages for age at oral emergence of the permanent lower
and upper first molars are between 5.2 and 5.9 years on Pima
Amerind children of Arizona; black children of Gambia, Ghana,
and Kenya; and Melanesian children of Bougainville, New Guinea,
and Siassi islands. For the same teeth, averages are between 6.0
and 6.5 years on Australian aborigines, children of Chinese ancestry
at Hong Kong, Sherpa highland children of Nepal, United States

black children, and white children residing in Australia, England, New Zealand, and the United States.

3. For the permanent upper central incisor teeth, averages are between 6.4 and 6.7 years on children of Bougainville and Siassi islands, New Guinea Kaiapit children, Polish children at Warsaw, and black children of Ghana and Kenya. Other averages are between 7.0 and 7.3 years on such diverse populations as Australian aboriginal children, Bundi children of New Guinea, children of Chinese descent at Hong Kong and on Taiwan, children of Japanese descent in Brazil and Japan, Mandinka village children in Gambia, Polish children of the Ostroleka and Suwalki districts, Sherpa highland children in Nepal, and white children in Australia, New Zealand, and the United States.

Averages for age at oral emergence of the permanent lateral incisor teeth are exhibited in Table 40. It can be seen:

1. The permanent lower lateral incisor teeth, on average, emerge between ages 6.4 and 7.0 years for Bougainville and Siassi girls and boys, Kaiapit girls and boys, Polish children of each sex at Warsaw, and black children of each sex in Ghana and Kenya. Corresponding averages for each sex are between 7.1 and 7.5 years on Chinese children at Hong Kong and on Taiwan, Japanese children at Nagasaki, Mandinka children in Gambia, white children in rural Poland, New Zealand, and the United States. The lower lateral incisor teeth emerge more than 1 year earlier in Kaiapit children of New Guinea than in Thai village children and Sherpa highland children.

2. Averages for age at oral emergence of the upper lateral incisor teeth are between 7.4 and 8.2 years on black boys of Ghana, Kenya, and the United States; Melanesian boys on Bougainville, New Guinea, and Siassi islands; and white boys in Australia, England, and urban Poland. On Chinese, Hungarian, Japanese, Mandinka, and Pima boys, averages for these teeth are between 8.5 and 8.7 years.

3. Assuming the statistics assembled in Table 40 provide a sufficient base for characterizing children worldwide, typical ages at which permanent lateral incisor teeth pierce the gingivae are near 7.1 years in the lower jaw of girls, 7.4 years in the lower jaw of boys, 7.9 years in the upper jaw of girls, and 8.3 years in the upper jaw of boys. It follows that the permanent lateral incisors, on the whole, emerge earlier in girls than boys, and earlier in the mandible than the maxilla.

Table 40

Average Age in Years at Oral Emergence of the Permanent Lateral Incisor Teeth in Children of Both Sexes Studied Since 1940

Group	Lower Jaw Girls	Lower Jaw Boys	Upper Jaw Girls	Upper Jaw Boys
Kaiapit, New Guinea	6.9	6.8	7.3	7.4
Bougainville islander	6.9	7.0	7.6	7.5
Polish, Warsaw	6.7	7.0	7.3	7.8
Siassi islander, near New Britain	6.7	6.8	7.3	7.9
Ghana black, Sunyani region	6.9	6.6	7.8	8.0
Digo Bantu, Kenya	6.4	6.9	7.7	8.1
United States black, Michigan	6.7	7.2	7.9	8.1
United States white, Iowa City	7.1	7.2	7.9	8.1
Australian white, New South Wales	6.9	7.3	7.8	8.2
English, Birmingham	7.0	7.3	7.8	8.2
Bundi, New Guinea	7.0	7.5	7.5	8.2
New Zealand white, Dunedin	7.2	7.4	7.9	8.3
United States white, Michigan	7.1	7.5	7.8	8.4
Polish, rural districts	7.3	7.5	7.9	8.4
Hungarian, northeast Hungary	7.2	7.9	8.0	8.5
Brazilian Japanese, São Paulo	6.9	7.2	8.1	8.6
Mandinka, Gambian villages	7.1	7.5	8.1	8.6
Taiwan Chinese	7.1	7.4	8.2	8.6
Japanese, Nagasaki	7.1	7.5	8.3	8.7
Hong Kong Chinese	7.2	7.5	8.3	8.7
Pima Amerind, Arizona	7.3	7.6	8.3	8.7
Sherpa, Nepal highlands	8.0	8.8	7.6	8.9
Thai, Bang Chan village	7.6	8.2	8.8	9.1

There are findings that indicate a secular trend toward earlier emergence of the permanent first molar and incisor teeth. Studies at German cities report averages for the lower and upper first molars changing from 7.0 and 7.4 years in 1894 to 5.8 and 5.9 years in 1950. Studies on United States black children report averages for the lower and upper central incisors showing decrease from 6.6 and 7.5 years in the 1930's to 5.9 and 6.8 years in the late 1960's. On United States white children of Dutch ancestry studied during the 1930's, average age at emergence of the upper lateral incisors is 9.0 years: a comparable average on United States white children studied in the late 1960's is 8.1 years.

Among present-day white children, individuals vary in time of gingival emergence of permanent teeth from ages 5 to 9 years for the first molars and central incisors, and from ages 5.5 to 11 years for the lateral incisors. Small numbers of Bundi and Mandinka children show emergence of first molar or incisor teeth between ages 4 and 5 years. At the other extreme, a few Chinese children show emergence at ages 11 to 12 years for 1 or more first molar teeth, and at ages 12 to 13 years for 1 or more lateral incisor teeth.

Occasionally a permanent tooth is congenitally missing. Among the permanent first molar and incisor teeth, an upper lateral incisor is the tooth most likely to be absent. In a sample of 6,700 United States white children, 1.7% lacked this tooth on one or both sides.

Oral emergence of permanent canine, premolar, and second molar teeth in late childhood. On Sherpa children living at highland villages in Nepal, oral emergence of permanent canine (cuspid) teeth occurs at average ages near 8.2 years for the lower and upper teeth of girls, and at 9.2 and 9.7 years for the lower and upper teeth of boys. On Bougainville islanders, average ages at oral emergence of the permanent lower canines are 8.9 years for girls, and 9.3 years for boys. Other averages for this tooth at ages between 9.0 and 9.5 years have been obtained on girls of African black, Australian white, British, Japanese, Polish urban, and United States black ethnic groups.

Averages for age at gingival emergence of the permanent upper and lower first and second premolar teeth are between 8.0 and 8.5 years on Sherpa girls, and between 8.9 and 9.3 years on Sherpa boys. On African black, Japanese, and Melanesian girls, also on Polish urban and Puerto Rican children of both sexes, averages from 9.0 to 9.5 years have been reported for age at emergence of the first premolar teeth in the upper jaw. Among most white populations, less than 20% of children show emergence of any premolar dental unit by age 9.5 years. Sometimes by age 9.5 years, particularly among Bundi children of New Guinea and Digo children of Kenya, 1 or more permanent second premolars, or permanent second molars have emerged. (The permanent premolars replace the primary molars, and the permanent second molars emerge directly behind the permanent first molars.)

Sex differences in emergence of permanent teeth are greatest for the lower canines; on average, these teeth emerge almost 1 year earlier in girls than boys. There are jaw differences in time of permanent tooth emergence: on average, premolar teeth emerge earlier in the upper jaw, and other teeth earlier in the lower jaw. Average emergence ages are practically alike for homologous teeth on the right and left sides of each dental arch. Many individual differences occur in emergence order of the permanent canine, premolar, and second molar teeth.

Number of permanent teeth present in the mouth at successive childhood ages. Listed in Table 41 are averages for number of permanent teeth in the mouths of girls and boys at annual ages from 6 years to 9 years. These averages show:

1. On white girls living in Hungary, New Zealand, Poland, Sweden, and the United States, oral emergence has occurred for 7 to 8 teeth at age 7 years, 10 to 11 teeth at age 8 years, and 13 to 14 teeth at age 9 years.

2. At ages 7, 8, and 9 years respectively, white boys have about one less emerged unit of the permanent dentition than white girls.

3. Black children are more advanced in oral emergence of permanent teeth than white children. On African Digo boys, numbers of permanent teeth visible in the mouth average near 6, 9, 11, and 14 at ages 6, 7, 8, and 9 respectively. On New Zealand white boys, corresponding numbers are 3, 6, 10, and 12 teeth.

4. At age 8 years, average numbers of orally emerged permanent teeth are about 12 for Hindu and Zulu boys, and less than 10 for Japanese and Thai boys.

At ages in middle and late childhood, individuals differ greatly in number of permanent teeth present in the mouth. Variations for white children age 7 years are from no oral emergence of permanent teeth to 18 permanent teeth having perforated the gums. At age 9 years, the spread within white populations is from 2 to 25 permanent teeth. Individual differences in oral emergence of permanent dental units among African black children age 9 years extend from 5 teeth to 28 teeth.

Among white children, the oral emergence of 18 permanent teeth by age 7 years, or 25 permanent teeth by age 9 years, occurs rarely. About 90% of white children have 1 to 6 emerged permanent teeth at age 6 years, 3 to 10 at age 7 years, 6 to 13 at age 8 years, and 10 to 17 at age 9 years.

Table 41

Average Number of Permanent Teeth Having Pierced the Gingivae at Successive Childhood Ages

Group	6 Years	7 Years	8 Years	9 Years
Girls				
Japanese, Japan	2.3	----	9.4*	------
Hungarian, northeast Hungary	----	7.1	10.2	12.7
United States white, Boston area	2.6	7.1	10.7	12.8
United States white, Cleveland	----	7.6	10.6	13.0
United States white, Hagerstown	----	7.1	10.2	13.3
New Zealand white, Dunedin	3.1	7.6	11.0	13.4
Polish, rural and urban	2.2	7.9	11.2	13.6
Swedish, rural and urban	----	7.6	10.7	13.9
United States white, Chicago	2.8	7.3	11.0	14.0
Hong Kong Chinese	3.8	7.2	11.0	14.7
Digo Bantu, Kenya	6.5	9.3	12.2	16.6
Zulu, Natal and Zululand	7.2	9.6	11.8	16.5
Boys				
Japanese, Japan	1.6	----	8.6*	------
Hungarian, northeast Hungary	----	6.1	9.2	11.9
New Zealand white, Dunedin	2.3	6.3	9.8	11.9
United States white, Cleveland	----	5.8	9.3	12.0
United States white, Hagerstown	----	6.1	9.3	12.1
United States white, Boston area	1.6	6.2	9.7	12.1
United States white, Chicago	2.5	6.3	9.7	12.3
Polish, rural and urban	2.1	6.1	10.6	12.4
Swedish, rural and urban	----	6.9	9.7	12.5
Hong Kong Chinese	3.1	6.3	9.7	13.0
Digo Bantu, Kenya	5.9	8.7	11.1	14.0
Zulu, Natal and Zululand	----	9.7	12.1	14.2
Hindu, Lahore and Madras	----	----	11.9	14.4

* At age 8 years, averages for number of orally emerged permanent teeth on Thai girls and boys at Bang Chan village are 9.0 and 7.8 respectively.

A report in 1837 on 700 English children age 9 years states: no child showed oral emergence of any permanent teeth other than the first molars and incisors; consequently, the maximum number of emerged permanent teeth was 12. Today, more than 10% of white children age 9 years have at least 16 emerged permanent teeth.

Suggested Readings

*Eveleth, P. B., and J. M. Tanner **Worldwide variation in human growth.** (International Biological Programme 8) Cambridge: Cambridge University Press, 1976.

Fanning, E. A., and T. Brown **Australian Dental Journal,** 1971, 16, 41-43.

Giles, N. B., V. B. Knott and H. V. Meredith **Angle Orthodontist,** 1963, 33, 195-206.

Houpt, M. I., S. Adu-Aryee and R. M. Grainger **American Journal of Orthodontics,** 1967, 53, 95-99.

Knott, V. B., and H. V. Meredith **Angle Orthodontist,** 1966, 36, 68-79.

Lee, M. M. C., W. D. Low and K. S. F. Chang **Archives of Oral Biology,** 1965, 10, 849-861.

Malcolm, L. A. **American Journal of Physical Anthropology,** 1969, 31, 39-52.

Chapter XIII

Growth Trends for Body Size and Form from Birth through Childhood

Earlier chapters on body size and form have dealt largely with the growth status of children at particular ages, and occasionally with growth progress during 1-year periods. This chapter presents a series of graphs illustrating growth progress during periods of infancy and childhood varying from 2 to 9 years.

Trends between birth and age 4 years for four measures of body size. Two graphs (Figs. 1 and 2) portray trends drawn to averages for body length, body weight, head girth, and chest girth. Each trend represents children measured between 1960 and 1970. From these graphs, it is found:

1. Throughout infancy and early childhood, there is continuous slowing of growth rates. All 10 curves depict this.

2. The curves for head girth and chest girth show average increase in these measures is greater during the first 6 months following birth than during the period between ages 6 months and 4 years.

3. The trends for body length and body weight depict greater increase during the first 18 months after birth than between ages 18 months and 4 years.

4. Slow, intermediate, and rapid increase in average body length during infancy and early childhood are shown by Bundi, Costa Rican, and Dutch children. Hindu, Spanish, and Dutch trends illustrate slow, intermediate, and rapid increase in average body weight.

5. Averages for head girth are about equal on contemporary groups of Italian (Grosseto province) children age 18 months and Hindu (Delhi) children age 4 years.

6. For Italian children, average chest girth at age 15 months is about 50% higher than at birth. A corresponding increase of 50% on birth size is attained about 2 years later by Hindu rural children residing in the vicinity of Hyderabad.

[123]

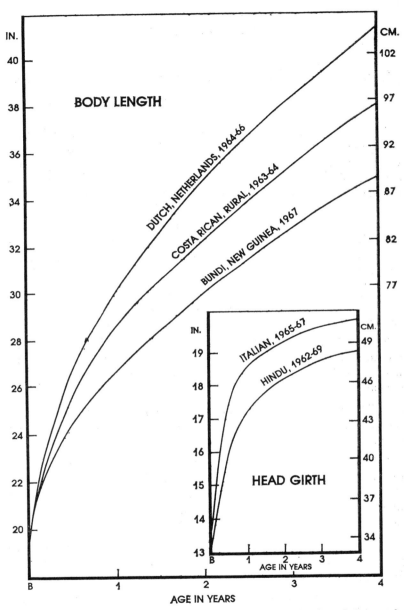

Fig. 1.   Trend lines drawn to averages at ages between birth and 4 years for body length and head girth.

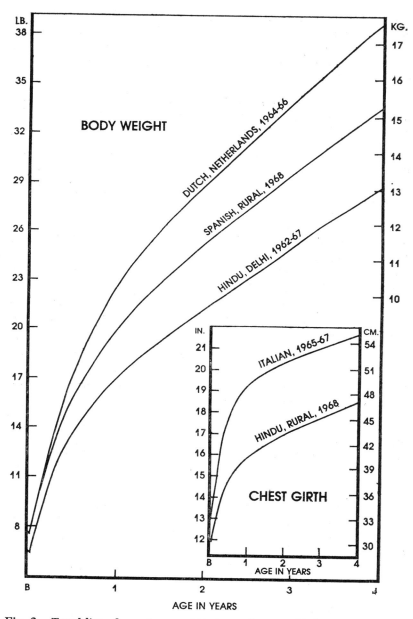

Fig. 2. Trend lines drawn to averages at ages between birth and 4 years for body weight and chest girth.

7. Averages for body length at birth are doubled by ages 3.5, 5.0, and 6.5 years on Dutch, Costa Rican, and Bundi children respectively. The averages for birth weight double before age 6 months, while those for head girth do not increase more than 65% between birth and adulthood.

8. In relation to averages for body weight of Dutch and Hindu children at birth, averages are four times higher at ages near 27 and 35 months respectively.

At age 2 years, Bundi children average about 50% of their height in early adulthood, and Dutch children about 50% (boys) and 53% (girls) of early adult height. Corresponding values are 19% (boys) and 21% (girls) for body weight of Dutch children, and near 88% for head girth of Italian children.

Trends in infancy and childhood for four indices of body form. In the upper part of Fig. 3 are trends for cephalo-thoracic and skelic indices typifying the course of change for each index throughout infancy and childhood. The lower part of Fig. 3 depicts trends extending between ages 3 and 9 years for hip width in percentage of lower limb height, and calf girth in percentage of lower limb height. Fig. 3 shows:

1. Throughout infancy and childhood, there is continuous increase in chest girth relative to head girth. There are group differences in rate of increase: the cephalo-thoracic index rises more slowly on Hindu children than Bulgarian children.

2. Average chest girth is near 104% of average head girth at ages near 3 years for Bulgarian children, and 6 years for Hindu children. At age 8 years, on average, chest girth relative to head girth is 106% on Hindu children, and 114% on Bulgarian children.

3. The skelic index increases at a slowing pace during infancy and childhood, with lower rates of increase for white than black children. At age 3 years, lower limb height of United States black children is about 77% of sitting height. This relationship is reached by Bulgarian children after age 5.5 years. Similarly, the skelic index is near 81 at age 4 years for United States black children, and age 6 years for Bulgarian children.

4. Hip width in percentage of lower limb height, and calf girth in percentage of lower limb height, each yield declining trends over the childhood period between ages 3 and 9 years.

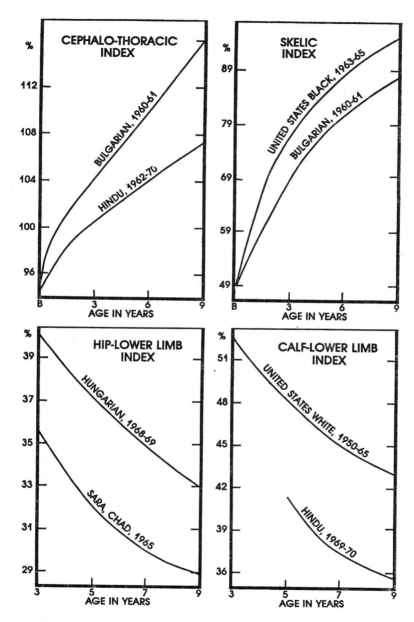

Fig. 3. Trend lines drawn to averages at ages between birth and 9 years for cephalo-thoracic and skelic indices, and at ages from 3 to 9 years for hip-lower limb and calf-lower limb indices.

5. On Sara black children, hip width relative to lower limb height decreases from near 36% at age 3 years to 29% at age 9 years. The comparable decrease for Hungarian white children is from near 40% to 33%.

6. The relatively narrow hips and long limbs of black children is registered in the comparison above, and on noting that indices are similar for Sara children age 4 years, and Hungarian children more than 4 years older.

7. Descent of the trend lines for calf girth in percentage of lower limb height reveals the manner in which, during childhood, the lower limbs become more slender with age. For United States white children, calf girth relative to lower limb height falls from about 52% at age 3 years, through 47% near age 6 years, to 43% at age 9 years. Hindu children, on average, have more slender lower limbs at age 5 years than United States white children at age 9 years.

Childhood trends for body size of racial, socio-economic, and secular groups. Standing height and body weight trends are presented in Fig. 4 on children of several racial groups studied from age 5 years to age 9 years. Findings contained in this graph are as follows:

1. Increase in average standing height is slightly greater between ages 5 and 7 years than between ages 7 and 9 years. Increase in average body weight is less in the biennium from 5 to 7 years than in the succeeding biennium.

2. Average standing height of Belgian children at Courtrai exceeds that of Lunda children in Angola by more than 4 inches (10 cm) at age 5 years, and 6 inches (15 cm) at age 9 years. At age 7 years, Chuvash children in the Soviet Union are over 2 inches taller than Lunda children, and over 2 inches shorter than Belgian children.

3. Belgian children age 6 years have about the same average height as Chuvash children 18 months older, and Lunda children 3 years older. Lunda children age 8 years are no taller than Belgian children age 5.5 years.

4. At age 7 years, average body weight is higher for New Zealand white children than Filipino children by more than 15 pounds (about 6.8 kg). At ages 5 and 9 years, the differences are about 12 and 19 pounds.

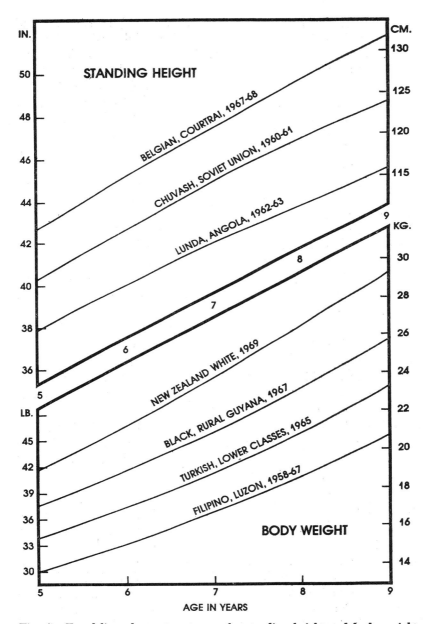

Fig. 4. Trend lines drawn to averages for standing height and body weight at ages from 5 to 9 years on contemporary black, white, Filipino, and Turkic-Tatar child populations.

5. Averages for body weight are near 42 pounds on New Zealand white children age 5 years, Guyana black children age 6 years, Istanbul children age 7 years living in Turkish homes of low socioeconomic status, and Filipino children a little older than 8 years.

Fig. 5 pertains to socio-economic differences. For Yoruba children at Ibadan, the children of underprivileged families are shorter than those of privileged families by more than 3 inches at each age between 6 and 9 years. At age 8 years, average body weight of the economically poor group is 10 pounds less than that of the well-to-do group.

The comparative trends for children of Chinese ancestry at Hong Kong show those in upper class homes surpass their age peers in lower class homes by about 1.5 inches in average height, and 2 to 3.5 pounds in average weight. As shown in chapter 9, differences between socio-economic subgroups are somewhat smaller for white populations of Europe and North America.

The trends in Fig. 6 portray secular differences in the body size of oriental, black, and white children. They demonstrate that at ages 7 to 9 years:

1. Japanese children living in 1900 were smaller than their age-race peers living 70 years later by 3 inches or more in average height, and about 1.5 inches in average chest girth.

2. English children living in 1910 were smaller than their age-race peers living 60 years later by 3 inches or more in average height, and 6 to 9 pounds in average weight.

3. United States black children living near the beginning of the present century were smaller than their age-race peers living 65 years later by 0.7 to 1.0 inch in average sitting height, and 1 to 2 inches in average lower limb height.

Charts for determining the status and progress of United States black and white children in height and weight. Figs. 7 and 8 are charts for assessing the height and weight status and progress of United States black and white children. Fig. 7 is adapted from charts constructed in 1977 for interpreting height and weight on United States black children and youths, and available through the College of Health and Physical Education, University of South Carolina, Columbia. Fig. 8 is adapted from similar charts constructed in 1975 for appraising United States white children and youths, and available through the American Medical Association, Chicago.

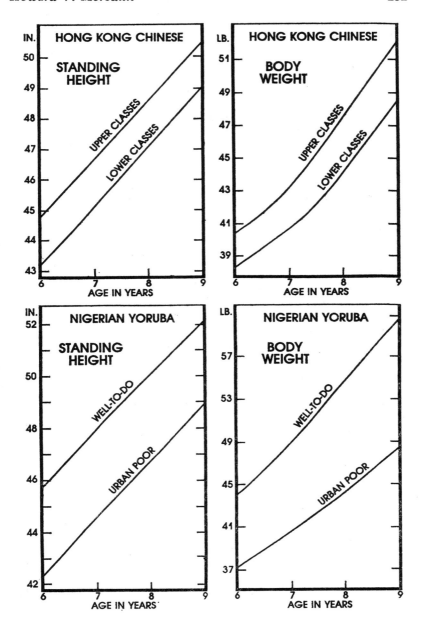

Fig. 5.   Trend lines drawn to averages for standing height and body weight
at ages from 6 to 9 years on children reared in homes of high and
low socio-economic status.

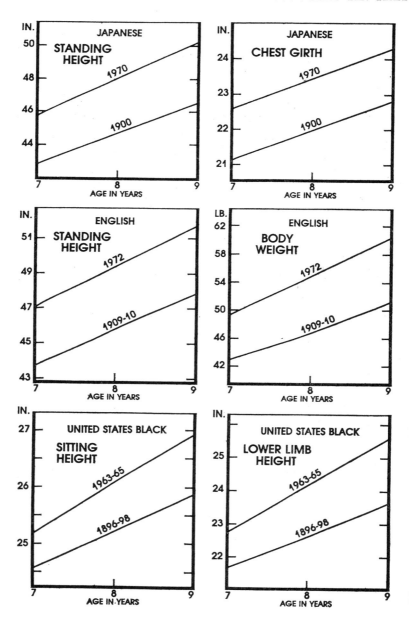

Fig. 6. Trend lines drawn to averages at ages 7, 8, and 9 years for standing height, body weight, chest girth, sitting height, and lower limb height of children living at times separated by 60 to 70 years.

The charts identified as Figs. 7 and 8 can be discussed together, regarding the one as appropriate for assessing black children, and the other for assessing white children.

Assume a black child's height has been measured at age 5 years, and found to fall in the Fig. 7 channel marked "tall." This child is among the upper 10% of United States black children age 5 years. Had the child's height fallen in the channel marked "short," he or she would have been among the lower 10% of United States black children age 5 years. Generalizing: the height portion of the chart is constructed to provide that 10% of black children place in the tall channel, 20% in the moderately tall channel, 40% in the average channel, 20% in the moderately short channel, and 10% in the short channel. The weight portion of the chart, and both portions of Fig. 8, are constructed in the same way.

By taking a child's height and weight at any age between 5 and 9 years, and plotting these records on the appropriate chart, height-weight status of the child can be determined. One child may be moderately short in height and average in weight, another tall in height and moderately heavy in weight, and so forth.

When a child's height and weight records do not fall in corresponding or adjacent channels, the discrepancy may reflect an undesirable state of health, or may depict normal slenderness or stockiness. Assume a child is found to be average in height and light in weight. This child should be referred for health examiners to determine whether he or she is a slender "satisfactorily healthy" child, or one with an incipient infection, a nutritional deficiency, or an unsuitable program of activities.

Having measured a child at a particular age, it is advantageous to measure again at later ages. Plotting a child's height and weight measurements at 6-month intervals gives a record of the child's progress in height and weight. During childhood, normality of growth progress is indicated by fairly parallel relationship of the individual's height and weight records with the channel lines of the chart. Marked deviations are grounds for thorough health appraisal. For example, a child's successive height records at ages from 5 to 8 years might fall along the middle of the channel marked moderately tall, while successive weight records fall near the middle of the moderately light channel from ages 5 to 8 years and then take a steep upward turn between ages 7 and 8 years. This child should be referred for study: the disproportionate gain in weight may denote need for a prescribed diet, a change in daily regimen, or drug therapy.

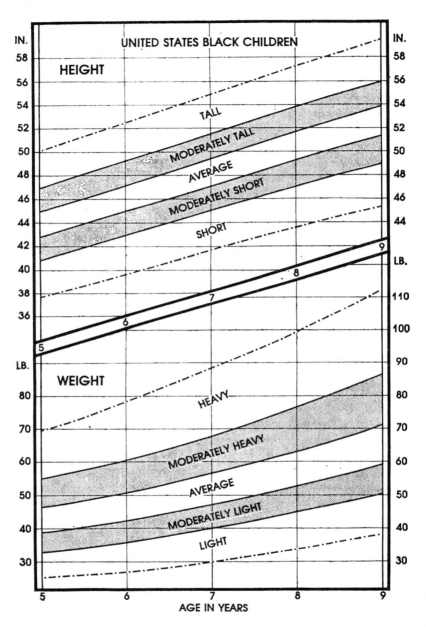

Fig. 7. Charts for assessing the height and weight status and progress of United States black children.

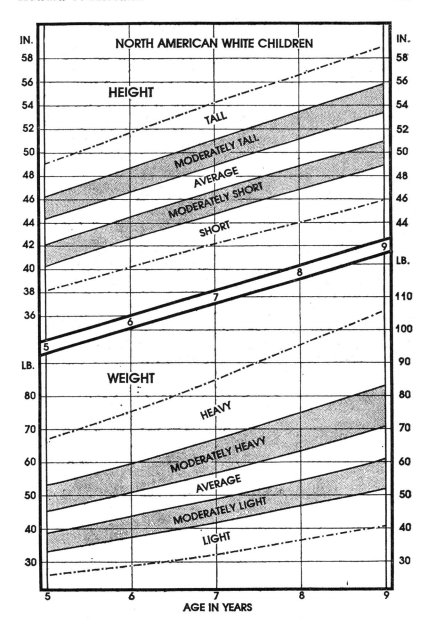

Fig. 8. Charts for assessing the height and weight status and progress of North American white children.

Suggested Readings

Bureau of Health Education **Height-weight interpretation folder for girls (boys).** Chicago: American Medical Association, 1975. (4-page booklet for each sex)

Hamill, P. V. V., T. A. Drizd, C. L. Johnson, R. B. Reed and A. F. Roche. **NCHS growth curves for children: birth-18 years, United States.** (DHEW Publ. No. PHS-78-1650, National Center for Health Statistics, Series 11, No. 165) Washington: Government Printing Office, 1977.

Meredith, H. V. **American Journal of Public Health,** 1949, 39, 878-885.

Meredith, H. V. **Journal of School Health,** 1955, 25, 267-273.

Spurgeon, J. H., and H. V. Meredith **Height and weight charts for Black-American children and youths of the United States.** Columbia: College of Health and Physical Education, 1977. (4-page booklet for each sex)

Tanner, J. M. **Archives of Disease in Childhood,** 1952, 27, 10-33.

Tanner, J. M., H. Goldstein and R. H. Whitehouse **Archives of Disease in Childhood,** 1970, 45, 755-762.

Postscript

To those who envision becoming future investigators of human body growth:

This book has dealt largely with research **findings** on body growth during the decade of human life beginning 9 months before birth. It has not included systematic discussions on (1) the concept of body growth and its major facets, (2) definitions and reliabilities of procedures used in obtaining growth data, (3) statistical methods appropriate for analyzing different sorts of data, or (4) steps in developing an original growth topic from library search for related materials, through research design and laboratory activities, to sound generalization. Cited below are a few introductory references pertaining to concept, method, and inference.

Bock, R. D., H. Wainer, A. Petersen, D. Thissen, J. Murray and A. Roche **Human Biology,** 1973, 45, 63-80.

Hammett, F. S. **The nature of growth.** Lancaster, Pa.: Science Press, 1936.

Krogman, W. M. **Growth of man.** Den Haag: Tabulae Biologicae, 1941.

Meredith, H. V. **University of Iowa Studies in Child Welfare,** 1943, 19, 1-337.

Meredith, H. V. **Physical Educator,** 1950, 7, 47-48.

Meredith, H. V. A descriptive concept of physical development, in D. B. Harris (Ed.), **The concept of development.** Minneapolis: University of Minnesota Press, 1957. (109-122)

Meredith, H. V. Methods of studying physical growth, in P. H. Mussen (Ed.), **Handbook of research methods in child development.** New York: John Wiley, 1960. (201-251)

Tanner, J. M. **Human Biology,** 1951, 23, 93-159.

Tanner, J. M., J. Hiernaux and S. Jarman Growth and physique studies, in J. S. Weiner and J. A. Lourie, **Human biology: a guide to field methods.** (International Biological Programme, No. 9) Oxford: Blackwell Scientific Publications, 1969. (1-76)

Subject Index